AN INTRODUCTION TO
AIR CHEMISTRY

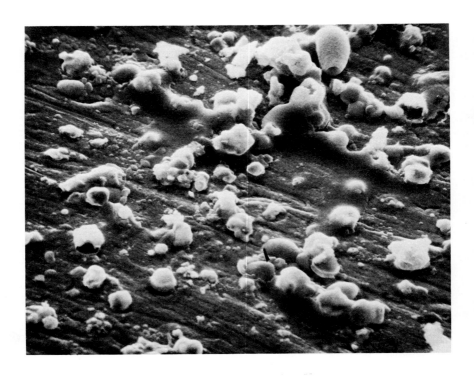

Atmospheric particulate matter on the surface of a stainless steel impactor plate as viewed with a scanning electron microscope operating at 1385× magnification. The area in view is ∼66 μm across. The impactor had a small size cutoff of ∼0.8 μm, and collected particles at the University of Washington in Seattle on August 26, 1971, 0807–0850 PDT. Sample and photo courtesy of Dennis Schuetzle.

AN INTRODUCTION TO
AIR CHEMISTRY

SAMUEL S. BUTCHER

Bowdoin College

ROBERT J. CHARLSON

University of Washington

ACADEMIC PRESS   New York and London

ACADEMIC PRESS, INC.
111 Fifth Avenue, New York, New York 10003

United Kingdom Edition published by
ACADEMIC PRESS, INC. (LONDON) LTD.
24/28 Oval Road, London NW1

LIBRARY OF CONGRESS CATALOG CARD NUMBER: 72-77348

PRINTED IN THE UNITED STATES OF AMERICA

To

Sally and Patricia

CONTENTS

PREFACE   xi

ACKNOWLEDGMENTS   xiii

Chapter 1   **Introduction**

1.1	Approach to a Problem	1
1.2	Evolution and Present Composition of the Atmosphere	2
1.3	Chemical Cycles	7
1.4	Physicochemical Models	10
	Problems	23
	References	23

Chapter 2   **Summary of Chemical Principles**

2.1	Properties of Gases	26
2.2	Properties of Solutions	27
2.3	Properties of Small Droplets and Particles	28
2.4	Chemical Thermodynamics	29
2.5	Extinction of Radiation	31
2.6	Chemical Kinetics	32
	Problems	36
	References	37

Chapter 3   **Sampling and Collection**

3.1   Generalized Methods of Analysis   38
3.2   Sampling Methods   40
3.3   Sampling Trains   52
      Problems   60
      References   60

Chapter 4   **Treatment of Data**

4.1   Instrumental Error   62
4.2   Presentation of Data   67
      Problems   74
      References   75

Chapter 5   **Special Methods of Analysis**

5.1   Chromatography   76
5.2   Spectrometry   82
5.3   Nuclear Methods   92
5.4   Mass Spectrometry   93
5.5   Remote Sensing Applications of Lasers   95
5.6   Correlation Spectrometry   96
      General References   97
      References   98

Chapter 6   **The Atmospheric Chemistry of Sulfur Compounds**

6.1   Global Considerations   100
6.2   Reactions of Sulfur Compounds   105
6.3   A Mechanism for the Oxidation of SO_2 106
6.4   Analytical Methods   108
      Problems   112
      References   113

Chapter 7   **Nitrogen Compounds and Ozone**

7.1   Reactions of Nitrogen Compounds   115
7.2   Global Aspects   122
7.3   Analytical Methods   123
      Problems   129
      References   131

Chapter 8   **Carbon Compounds**

8.1	Sources and Sinks of Carbon Dioxide	134
8.2	Sources and Sinks of Carbon Monoxide	139
8.3	Behavior of Carbon Oxides in the Atmosphere	141
8.4	Other Carbon Compounds	142
8.5	Analytical Methods	150
	Problems	154
	References	155

Chapter 9   **Aerosols**

9.1	Introduction	157
9.2	Sources and Sinks of Particulate Matter	158
9.3	Particle Size Distribution	165
9.4	Mechanical Properties of Aerosols	176
9.5	Diffusion and Coagulation	181
9.6	Optics of Aerosols	184
9.7	Measurement on Aerosols	192
	Problems	207
	References	208

Appendix 1   **Units and Dimensions Used in Air Chemistry**

211

Appendix 2   **List of Symbols**

214

Appendix 3   **Glossary**

219

AUTHOR INDEX	231
SUBJECT INDEX	236

PREFACE

This book is designed to satisfy the increasing need for a textbook on the subject of air chemistry for those with no previous experience in the field. Although there are a number of books which treat individual aspects of this problem, the areas of analytical chemistry and meteorology have not been brought together in a book designed as a textbook. We hope to achieve the synthesis of chemistry and meteorology required to bring into focus some of the problems associated with our atmospheric environment. We have also attempted to describe an approach to a complex problem in a manner which will endure even after some of the specific analytical techniques have changed. The philosophy of the book is also summarized in the first few pages of Chapter 1.

The prerequisites for a course in which this text might be used would include a year each of college chemistry and calculus. Although students interested in this field are encouraged to obtain a much more extensive preparation in the physical sciences, such preparation is not required for the use of this book. At present, courses in which a book of this nature might be used are offered as upper-level undergraduate and graduate-level courses in departments of meteorology, oceanography, civil engineering, or in those university departments in which there is activity in the field of air resources.

In addition to its use in introductory courses on air chemistry, this book should also serve as a reference book for practicing technical people in the air pollution field. Laboratory chemists, meteorologists, technicians, and those working in field sampling of air pollutants can use it to supplement their training. Chemists, meteorologists, and physical scientists in general who wish an introduction to the air chemistry field should find it useful.

The reader is introduced to the overall subject and given a review of the relevant chemical and meteorological principles in Chapters 1 and 2. Since space is not available to include many details, references are included to assist the reader whose background is deficient. Chapters 3–5 discuss the general methods of obtaining and handling air chemical data. Chapters 3 and 4 are quite general; Chapter 5 introduces the student who has not had a course in modern analytical chemistry to some of the analytical methods available. Three classes of chemical compounds which are important in any consideration of trace constituents of the atmosphere are discussed in Chapters 6–8. Significant atmospheric reactions, the global budgets, and selected methods of analysis for these compounds are considered. The final chapter treats some of the physical characteristics of aerosols, a subject which is very important for any consideration of atmospheric chemistry, but which is not usually included in courses in the field of chemistry or meteorology.

So that the reader is not misled, we want to emphasize that this book has a finite and limited goal of presenting *basic* air chemistry information. While some reference is made to current research, it cannot be considered as a research-oriented text, but rather as a utilitarian work that will introduce the reader to successful approaches to the study of air chemistry.

ACKNOWLEDGMENTS

This book has an unusual history in that one of us (S.S.B.) selected (almost by chance) the University of Washington for sabbatical study, thus bringing his interest in chemistry to the physical/meteorological area of the other (R.J.C.). The book evolved from a set of class notes which had grown over a period of years. Professors R. F. Christman and M. J. Pilat have taught the course (besides R.J.C.) and we acknowledge their contributions to the subject matter. Comments and suggestions on the manuscript by Professors F. I. Badgley of the Atmospheric Sciences Department and A. W. Fairhall of the Department of Chemistry, University of Washington are gratefully acknowledged, as is the substantive review of Chapter 9 by Dr. Marcia Baker. Thanks are due to Ms. Pam Labbe and Ms. Lin Ahlquist for their patience in preparing the final typescript and the figures. We also want to thank the many students of the Water and Air Resources Division of the Civil Engineering Department, the Atmospheric Sciences Department, and elsewhere, who patiently studied without the benefit of a textbook.

CHAPTER 1

INTRODUCTION

1.1 Approach to a Problem

Information regarding the chemistry of atmospheric air has been developed in several technical fields over the past decade or two. This information is diverse in character, ranging from analytical techniques for urban pollutants to methods for remotely sensing upper atmospheric constituents from satellites. Common substances such as SO_2, CO, and other pollutant molecules have been studied along with esoteric substances generated by a wide variety of natural and anthropogenic sources. As a result of the breadth of the subject, there are few places where information is collected and organized.

The main purpose of this book is to provide basic and useful information on atmospheric chemistry. Since it is not possible (and probably not desirable) to include all of this vast subject in one book, there are several limitations which have been imposed. First, the information is *basic* to a wide variety of problems in the chemistry of air. Second, no attempt is made to include the fringes of the subject by describing all current research. Third, data and methods have been selected because they are useful or because they provide useful insight into the subject. No attempt is made to be exhaustive in literature review nor to catalog all analytical methods. Rather, care was taken to provide an adequate variety of analytical methods so that the nature of successful approaches can be demonstrated.

A considerable part of this text is devoted to methods of sensing or measuring substances in air. While it is possible to utilize these methods directly in field or laboratory studies, it should be emphasized that other and newer methods *always* will exist and that it is particularly hazardous to treat any text as a "cookbook."

The reader should rather study the methods in order to gain insight into the approach and into the limitations of various approaches to a measurement. For example, wet chemical methods for trace gas analysis are presented, as are physical methods. Each has its advantages and disadvantages. More importantly, the aggregate of methods provides a variety of chemical approaches to the measurement of trace gases. This breadth will hopefully provide the insight needed for approaching the novel situation not covered in any "cookbook." This treatise is not intended as a contradiction of the "standard method" approaches of American Society for Testing and Materials, but rather should provide an understanding of their bases, how to use them, and even how to modify or improve them.

Much of the lack of organized information on atmospheric chemistry can be traced to the single fact that there are very few people in the United States (or the world, for that matter) who identify themselves as "atmospheric chemists." Inasmuch as such a group could ever be identified, they would probably first be recognized by the way that they approach problems in air chemistry. Their repertoire of subject matter and the way they use it would be the hallmark of their field.

While it is not the purpose of this book to establish a new field, it is clearly important to provide a fledgling discipline some support. In particular, the point of view herein presented is that of the atmospheric chemist. The tools of chemistry and meteorology alone do not suffice to provide a realistic approach to this subject. The broader, interdisciplinary view herein discussed has proven to be realistic and useful. It is hoped that it will also aid in identifying atmospheric chemistry as a useful and functional part of science.

1.2 Evolution and Present Composition of the Atmosphere

About five billion years ago, the earth was formed, probably out of dust and gases in space. Little is known about the exact sequence of these formation processes, although we can deduce that the early earth did have an atmosphere at a point in time when gravity was sufficient to prevent atoms and molecules from escaping.

Our knowledge of the approximate abundance of elements in our solar system suggests that the early atmosphere was largely H_2, He, CO_2, CH_4, and H_2O, with traces of N_2 and noble gases. However, that abundance changed over the eons to the present composition.

One of the major changes involved a decrease by several orders of magnitude in the amounts of native noble gases, ^{36}Ar and ^{38}Ar, Xe, Kr, and Ne (i.e., those not produced by radioactive decay processes, like ^{40}Ar). This deficiency might have been present originally, but the most popular theories attribute it to a thermal event in which the atmospheric gases were given escape energies by high temperature, explosion, or other cataclysmic events. Escape from the earth's gravity requires a velocity of about 11 km/sec, higher than those provided even by large volcanoes and nuclear explosions.

Another major change was the appearance of large amounts of CO_2 and H_2O vapor, which subsequently were removed as atmospheric gases. Rubey (1951) suggested the source of these as "excess volatiles" from the original formation of rocks in the earth's crust and interior. Hot springs and volcanoes are present-day reminders of the kinds of venting that could have occurred billions of years ago on a grander scale. The H_2O vapor condensed into the oceans and hydrosphere, while CO_2 became involved in life processes and was deposited as limestone ($CaCO_3$). Some H_2O was photolyzed to form H_2 and O_2.

Meanwhile, lighter gases such as He and H_2 were lost, and are still being lost from the upper reaches of the atmosphere by diffusion and by the high-energy "tail" of the Maxwell velocity distribution, which provides a small fraction of the gas with velocities greater than the escape velocity.

Thus little of the primitive atmosphere remains. The mass of material—largely CO_2—that has been lost to the lithosphere in the formation of carbonates in limestone is such that the early atmosphere could have been very dense—perhaps similar to the current atmosphere of Venus with a surface pressure 100 times our own.

Few theories explain in detail the evolution of our present N_2, which accounts for about 80% of the mass of the atmosphere. The best guess seems to be that it is itself one of the primitive "excess volatiles," and that it is sufficiently unreactive to have stayed in elemental form and sufficiently high in molecular weight (28) not to be lost by diffusion.

Perhaps the most interesting major atmospheric constituent is the highly reactive material O_2. If the earth ever existed as a hot body, the

TABLE 1.1

AVERAGE GASEOUS COMPOSITION OF DRIED AIR

Symbol	Name	$\bar{c}$, Average concentration, volume fraction	$(1/\bar{c})\,dc/dt$ Order-of-magnitude estimate	Comments	Ref.
N_2	Nitrogen	0.78084	Very small	—	a
O_2	Oxygen	$0.20946 \pm 6 \times 10^{-5}$	-5×10^{-6} year^{-1}	Maximum decrease based on measurements	b
^{40}Ar	Argon	9.34×10^{-3}	—	—	a
CO_2	Carbon dioxide	3.25×10^{-4}	$+3 \times 10^{-3}$ year^{-1}	Decadic average, nonurban	c
		3.25×10^{-4}	$\pm 1 \times 10^{-2}$ year^{-1}	Normal annual variation, nonurban	c
		$3.25\text{–}10 \times 10^{-4}$	± 1 hr^{-1} or more	Urban	d
Ne	Neon	1.818×10^{-5}	—	—	a
He	Helium	5.24×10^{-6}	—	—	a
CH_4	Methane	$1.2\text{–}2.0 \times 10^{-6}$	± 1 day^{-1}	Few measurements	e

Kr	Krypton	1.14×10^{-6}	—		a
H$_2$	Hydrogen	5×10^{-7}	—	Few measurements	a
Xe	Xenon	8.7×10^{-8}	—	—	a
CO	Carbon monoxide	8×10^{-8}–5×10^{-7}	$\pm 1\ \text{day}^{-1}$	Nonurban	e
		10^{-6}–10^{-4}	$\pm 1\ \text{hr}^{-1}$ or more	Urban	f
N$_2$O	Nitrous oxide	2–4×10^{-7}	$\pm 0.1\ \text{hr}^{-1}$	Tropical and temperate, nonurban	g
SO$_2$	Sulfur dioxide	7×10^{-9} (rural) to $>1 \times 10^{-6}$ (urban)	$\pm 1\ \text{hr}^{-1}$ or more	Extremely variable	h
NO	Nitric oxide	10^{-8}–10^{-6}	$\pm 1\ \text{hr}^{-1}$ or more	Urban pollutant	h
NO$_2$	Nitrogen dioxide	10^{-8}–10^{-6}	$\pm 1\ \text{hr}^{-1}$ or more	—	h
HCHO	Formaldehyde	$\leq 10^{-7}$	$\pm 1\ \text{hr}^{-1}$ or more	—	f
NH$_3$	Ammonia	$\leq 10^{-6}$	$\pm 1\ \text{hr}^{-1}$ or more	Few measurements	i
O$_3$	Ozone	0–5×10^{-8} (rural) to 5×10^{-7} (urban)	$\pm 1\ \text{hr}^{-1}$ or more	Near sea level	d

[a] U.S. Standard Atmosphere (1962).
[b] Machta and Hughes (1970).
[c] SMIC (1971).
[d] Authors' measurements.
[e] Swinnerton et al. (1969).
[f] Chapter 8.
[g] La Hue et al. (1970).
[h] Various.
[i] Healy et al. (1970).

O_2 must have been bound up as oxides. Much thought has been given to the evolution of terrestrial O_2, which in our solar system is apparently unique to the earth. The most popular explanation is that of Berkner and Marshall (1964), who attribute essentially all of our present O_2 to the biospheric process of photosynthesis. They suggest that only the first necessary traces of O_2 came from H_2O photolysis. Some disagreement exists with this hyphothesis, however, based on a number of other possibilities (Van Valen, 1971).

This thumbnail history of the atmosphere leads us to two points: that the atmosphere has a constantly changing chemical composition, and that any average composition we presently define is likely to change on a geologic time scale. It is important to realize that the natural rates of change of *most* natural atmospheric constituents are very small. If all the O_2 in the atmosphere were produced in the period of 10^8–10^9 years, the change in the last 100 years (at the same rate), when man could contemplate measuring O_2, would be 10^{-6}–10^{-7} of the present levels. The present concentration of O_2 is 20.946% by volume, with an uncertainty of about 0.006% (Machta and Hughes, 1970). The same authors have also shown that the change in O_2 concentration in the last 60 years is less than 0.010%, suggesting that it is practical to consider O_2 to be fixed.

There are other natural atmospheric gases with more rapid rates of change. A few degrees change in the temperature of the oceans could change the CO_2 content by many per cent of its present level ($\sim$0.03% of the total). Water vapor content changes on a very short time scale, and on a small spatial scale as well.

By far most of the short-term variation of composition occurs with substances of very low concentration. As a result, the study of atmospheric chemistry focuses on such "trace" substances since fewer interesting changes occur in the N_2, O_2, and ^{40}Ar contents.

Table 1.1 provides data on the average composition of dry air and of the typical rates of change of the constituents.

A useful quantity to use as an index of variability is $(1/c)\, dc/dt$. Some values for this quantity have been estimated from recorded or published data. Of particular interest is oxygen (O_2), which in the recent past has attracted attention based on possible future changes. Both man's consumption of oxygen by burning fuels and the decrease in photosynthetic activity as a result of deforestation, water pollution (killing phytoplankton), etc. have been listed as causes for alarm. Broecker (1970) showed, however, that this concern is indeed unwarranted due to the large mass of O_2 in the atmosphere. The value of $(1/c)\, dc/dt$ is indeed small for global oxygen, and is even very small in cities where carbon based fuels are consumed.

It is significant and interesting that the variability $(1/c)\,dc/dt$ of atmospheric constituents falls into two categories:

(a) Stable or "fixed" constituents, with

$$10^{-6}\ (\text{or less}) < |\ (1/c)\ dc/dt\ | < 10^{-2}\quad \text{hr}^{-1}$$

(b) Variable constituents, with

$$10^{-1} < |\ (1/c)\ dc/dt\ | < 10\quad \text{hr}^{-1}$$

While overlap of these magnitudes can occur, it seems to be rare. Thus, the interest of atmospheric chemists focuses on the latter groups of substances since little variability can be observed in the former. This variability, its causes, and effects are a key issue in this subject.

It would be possible to generate another table like Table 1.1 but with $(1/c)\,dc/dx$ as the variability index (where x is a spatial coordinate, such as the vertical axis). Similar results would occur in these groupings. In particular, the historical interest of scientists in the variability of upper atmospheric O_3 would be indicated. The role of O_3 in heating the stratosphere is the main reason for upper atmospheric O_3 studies.

1.3 Chemical Cycles

On any scale (i.e., physical dimension), the atmosphere exhibits *both* physical and chemical variations of significant magnitude. Very often, the rates of change of concentration of a substance can be described as being due to both chemical reactions and physical diffusion processes. At this time, we do not have a firm understanding of even the number of important processes, let alone a complete grasp of their quantitative importance.

From the preceding discussion, we do know that, with few exceptions, the time rates of change on the global scale are very small—that is, we are not accumulating or losing atmospheric constituents at a rapid rate. This leads to the concept of an approximately balanced budget of the chemical ingredients of the atmosphere. These budgets can be analyzed in terms of the cycles of individual elements which, taken together, provide a useful view of the whole atmosphere. Figures 1.1a and 1.1b represent the aggregate of these elemental cycles depicting the overall atmosphere and its stable chemical behavior. The details of some of the individual cycles will be discussed later in the context of the discussion of the chemistry of the specific substances. Here, we want to gain an overview to demonstrate the complexity and dynamics of the whole system.

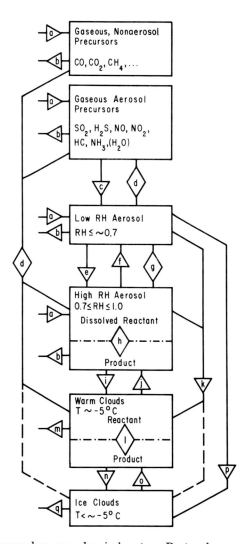

FIG. 1.1a. The troposphere as a chemical system. Rectangles are recognizable entities in the atmosphere. Triangles represent processes which have a single direction of material flow, and diamonds (two triangles) represent reversible processes. (a) Sources. (b) Sinks. (c) Gas-to-particle conversion. (d) Sorption. (e) Deliquescence. (f) Efflorescence. (g) Raoult's equilibrium. (h) Reaction in concentrated solution droplet. (i) Nucleation and condensation of water. (j) Evaporation. (k) Capture of aerosol by cloud drops. (l) Reaction in dilute solution. (m) Rain. (n) Freezing of supercooled drop by ice nucleus. (o) Melting. (p) Direct sublimation of ice on ice nucleus. (q) Precipitation. Note: This figure and Fig. 1b are drawn separately for convenience. It must be realized that exchanges between the stratosphere and troposphere also occur.

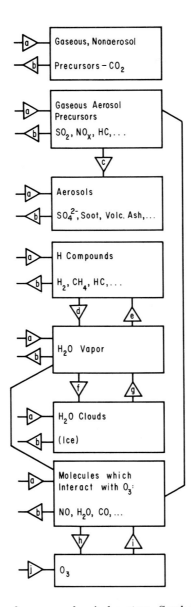

FIG. 1.1b. The stratosphere as a chemical system. Symbols convey the same significance as in Fig. 1a, with the following specific processes: (a) Sources. (b) Sinks. (c) Gas-to-particle conversion. (d) Oxidation reactions. (e) Reduction reactions. (f) Nucleation and deposition of water. (g) Sublimation. (h) Ozone-producing reactions. (i) Ozone-consuming reactions. (j) Radiation.

1.4 Physicochemical Models

If we take air samples from the atmosphere for subsequent chemical analysis, we find that the minor constituent composition exhibits complicated variations. Many of these are natural, but most of the variations demanding our attention are due to human production of pollutants. In either case, we can see readily that the chemical composition is dependent on all the normal spatial variables and time, including the time dependence of the sources and sinks of the substance of concern. Further, variations in composition can be noted on spatial scales from meters to thousands of kilometers, with some substances requiring different spatial resolution than others. Figure 1.2 illustrates this complexity.

Another way of describing this complex situation is to use a generalized differential equation describing the rate of change of concentration (Friend and Charlson, 1969):

$$dc/dt = Q + R_{prod} - R_{rem} - D - A - P \qquad (1.1)$$

This says that the concentration in an air volume changes at a rate dc/dt that is a function of the magnitude of the sources contributing material to the volume, Q; production and removal reactions within the volume, R; the loss of material by diffusion and dispersion, D; loss by advection, A; and loss by precipitation, P. In either case, it is clear that large amounts of information must be available for the total quantitative description of the dynamics of the cycles involved. Such a quantitative description is required

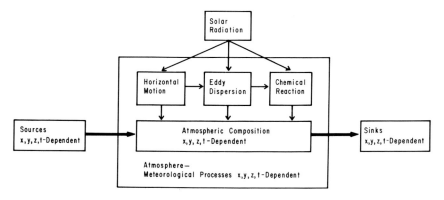

FIG. 1.2. The simultaneous interaction of sources, meteorological processes, chemical reactions, and sinks which determines the air composition at one point in space and time. The wide arrows represent material flow, the narrow arrows indicate the main direction of effects of the given processes.

if accurate knowledge of the relationship of source strength to atmospheric composition is to be obtained, for instance, for air pollution control strategy. For some substances (e.g., CO_2) it appears possible to utilize only one carefully chosen site for monitoring the whole atmosphere for long-term secular changes, while in other cases, the variations of a few city blocks in location or a few meters in sampling height can introduce gross changes in apparent behavior of a substance (e.g., CO).

In view of this kind of complexity, atmospheric scientists in other areas—such as energy balance—resort to the judicious use of data in establishing the boundary conditions for their models. There is no possibility of a hard and fast rule for determining the spatial and temporal resolution required in the taking of data. Likewise, there is no possibility for defining the residence time or reaction rate of a material injected into the atmosphere without adequate ancillary information on the general chemical situation.

Figures 1.1 and 1.2 showed that both chemical reactions and physical/ meterological processes influence the composition of air. While we are unable to model all of the chemical reactions with certainty, meteorologists have provided much useful information on those physical processes that concern us. *Diffusion, eddy dispersion, advection,* and *cloud* and *removal* processes are especially important, though they will be treated only briefly here. References to more complete works are given for those cases that require fuller study or explanation.

1.4.1   MOLECULAR DIFFUSION

Diffusion in the atmosphere proceeds by two different processes. *Molecular diffusion,* as described by Fick's law, results from a concentration gradient which causes molecules to move in the direction of lower concentration. The flux density F is given by Fick's law in a one-dimensional case as

$$F = -D\, dc/dx \tag{1.2}$$

or in three dimensions as

$$\mathbf{F} = -D\,\nabla c \tag{1.3}$$

In these equations, c is the concentration, x the spatial coordinate, and D the diffusion coefficient. If we consider mass continuity, a form of Laplace's equation results:

$$\partial c/\partial t = \nabla \cdot (D\,\nabla c) \tag{1.4}$$

TABLE 1.2

DIFFUSION COEFFICIENTS
IN AIR AT 0°C, 1 ATM[a]

	cm²/sec
H_2	0.634
H_2O	0.250
O_2	0.178
CO_2	0.139
$(C_2H_5)_2O$	0.099

[a] Weast, R. C. ed. (1968). "Handbook of Chemistry and Physics," 49th ed. Chem. Rubber Co., Cleveland, Ohio.

or, if D is independent of space,

$$\partial c / \partial t = D \, \nabla^2 c \qquad (1.5)$$

Table 1.2 provides data on the molecular diffusion coefficients for a few gases in air.

1.4.2   EDDY DISPERSION

Molecular diffusion is a relatively slow process compared with other atmospheric processes. It is effective over spatial scales of centimeters or less, but quite ineffective over larger scales except above about 120 km altitude. *Eddy dispersion* is a process which is effective in the atmosphere in the spatial scale of centimeters to kilometers or more. Eddy dispersion occurs when air is mixed by *turbulence* caused by shear, by flow over and around obstacles, including surface roughness, or by thermally produced buoyancy. The flux in this case is caused by individual fluid elements or "eddies" transporting the material. A flux occurs toward the region of lower concentration, just as in the case of molecular diffusion, except that the mechanism is different.

Flux density due to fluid motion is given by the product of concentration and velocity:

$$\boldsymbol{F} = c \boldsymbol{U} \qquad (1.6)$$

where $\boldsymbol{F}$ is the flux vector and $\boldsymbol{U}$ is the velocity vector. If the fluid is moving without turbulence, this equation describes the flux of material across a plane perpendicular to the flow.

If, on the other hand, there is turbulence—such as almost always exists in the atmosphere—further description is needed. The velocity vector can be thought of as the sum of an average velocity and a perturbation:

$$U = \bar{U} + U'$$

The concentration may also be represented as the sum of an average and a perturbation:

$$c = \bar{c} + c'$$

The product thus becomes

$$F = (\bar{U} + U')(\bar{c} + c')$$

$$= \bar{U}\bar{c} + \bar{U}c' + U'\bar{c} + U'c'$$

The average flux density $\bar{F}$ is simplified because the averages of the perturbation quantities are zero:

$$\bar{F} = \bar{U}\bar{c} + \overline{U'\bar{c}} + \overline{\bar{U}c'} + \overline{U'c'}$$

$$= \bar{U}\bar{c} + \overline{U'c'} \tag{1.7}$$

These two remaining terms describe the physically separable processes of *advection* and *eddy dispersion*, respectively. The latter provides a positive contribution to F when U' and c' have like signs, and subtracts from F when the signs are opposite, as shown in Fig. 1.3.

While this description is conceptually correct, it has not been generally possible to evaluate directly the quantity $\overline{U'c'}$ in many atmospheric cases. It is quite difficult to obtain simultaneous measurements of the velocity and concentration perturbations. It has usually been possible to evaluate the advection quantity.

A more empirical but practical approach has generally been used which depends upon an analogy between the random character of eddy motion and molecular (Brownian) motion (Fleagle and Businger, 1963, p. 186). Theoretical and experimental justification for such an approach has formed a large body of knowledge in meteorology (Slade, 1968, pp. 80–99).

Several different approaches (with varying assumptions) lead to a simple description of eddy flux density F across a plane normal to the x axis:

$$F = -K \, dc/dx \tag{1.8}$$

and to eddy dispersion (K assumed constant with x), where $\bar{c}$ is the time averaged concentration at a given point:

$$\partial \bar{c}/\partial t = K \, \partial^2 \bar{c}/\partial x^2 \tag{1.9}$$

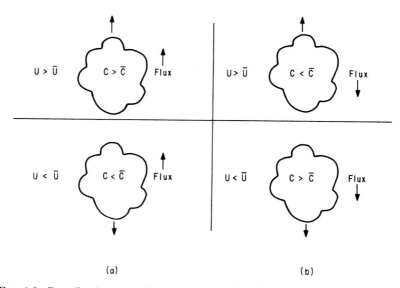

FIG. 1.3. Contributions to eddy flux density. The eddy is shown as a blob or puff with velocity $U = \bar{U} + U'$ and concentration $c = \bar{c} + c'$. If the product $U'c'$ is greater than zero, as in the two cases (a) on the left, an upward flux results. Otherwise, the flux is downward (b).

The solutions to such equations are well known in mathematics and physics as the analogs of the heat conduction equation. The fundamental solutions are of the form of Gaussian functions. Let us first consider the one-dimensional case of an instantaneous release of material from a point source.

If $\bar{c} \to 0$ as $t \to \infty$ $(-\infty < x < \infty)$ and $\bar{c} \to 0$ as $t \to 0$, except at $x = 0$, where $\bar{c}$ is assumed to approach infinity, the conservation-of-mass condition may be written

$$\int_{-\infty}^{\infty} \bar{c}\, dx = Q$$

where Q is the amount released at $t = 0$.
The solution is

$$\bar{c}/Q = (1/at^{1/2}) \exp(-bx^2/t) \tag{1.10}$$

It is easy to show from the original Laplacian equation that the coefficients a and b can be related to K, the eddy dispersion coefficient, sometimes called eddy diffusivity,

$$a = (4\pi K)^{1/2}, \qquad b = 1/4K$$

so

$$\bar{c}/Q = [1/(4\pi Kt)^{1/2}] \exp(-x^2/4Kt) \tag{1.11}$$

Another approach which allows for K to be dependent on the spatial coordinate results in an analogous solution for *nonisotropic* dispersion:

$$\frac{\bar{c}(x, y, z, t)}{Q} = (4\pi t)^{-3/2}(K_x K_y K_z)^{-1/2} \ \exp\left[-\frac{1}{4t}\left(\frac{x^2}{K_x} + \frac{y^2}{K_y} + \frac{z^2}{K_z}\right)\right] \tag{1.12}$$

This equation can also be integrated with respect to time to give results applicable to continuous point sources. If we think of the above as what happens to a puff of smoke emitted at time $t = 0$ and if there is a wind of horizontal velocity $\bar{U}$, we could think of the equations as describing events in a moving coordinate system. If we consider a constant source as emitting a series of finite puffs as in Fig. 1.4, then the concentration resulting at some distance can be calculated. Integration of this equation could then yield a realistic solution.

A major simplification is realized if dispersion in the x direction is neglected. Figure 1.4 simplifies to Fig. 1.5. Here, dispersion occurs across the plume, but downwind dispersion is assumed to have negligible effects. In other words, this says that the wind velocity is much greater than the dispersion velocity.

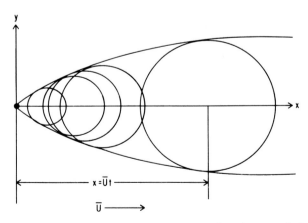

FIG. 1.4. A continuous source as a series of puffs undergoing translation in one direction, x, as well as one-dimensional dispersion from the translated origin. Here, $\bar{U}$ is the mean wind which translates the puff a distance $\bar{U}t$ in time t. The concentration at a given point would be the sum of contributions from all puffs at that point.

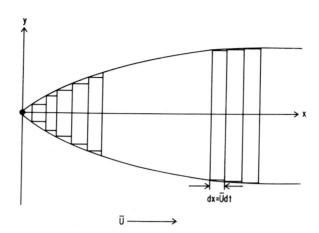

Fig. 1.5.   Plume from a continuous source with dispersion in the x direction assumed to be negligible.

Since the values of K_x, K_y, and K_z are determined experimentally for an appropriate set of atmospheric and topographic conditions, the equation can be further simplified as was suggested by Pasquill (Slade, 1968, pp. 99, 101) for a ground-level source:

$$\frac{\bar{c}(x, y, z)}{Q'} = (\pi\sigma_y\sigma_z\bar{U})^{-1} \exp\left[-\left(\frac{y^2}{2\sigma_y{}^2} + \frac{z^2}{2\sigma_z{}^2}\right)\right] \qquad (1.13)$$

Here, Q' represents the emission rate, and σ_y and σ_z the standard deviations of the Gaussian distribution. The other constants in the equation put it into standard Gaussian form. The quantities σ_y and σ_z have the dimensions of y and z, respectively, and may be thought of as a *scale* of plume width and depth, respectively.

The equation can be adapted to elevated sources of effective height h by translation:

$$\frac{\bar{c}}{Q'} = (2\pi\sigma_y\sigma_z\bar{U})^{-1}\left(\exp -\frac{y^2}{2\sigma_y{}^2}\right)$$

$$\left\{\exp\left[-\frac{(z-h)^2}{2\sigma_z{}^2}\right] + \exp\left[-\frac{(z+h)^2}{2\sigma_z{}^2}\right]\right\} \qquad (1.14)$$

or if $z = 0$ (i.e., ground level),

$$\frac{\bar{c}}{Q'} = (\pi\sigma_y\sigma_z\bar{U})^{-1} \exp\left[-\left(\frac{y^2}{2\sigma_y{}^2} + \frac{h^2}{2\sigma_z{}^2}\right)\right] \qquad (1.15)$$

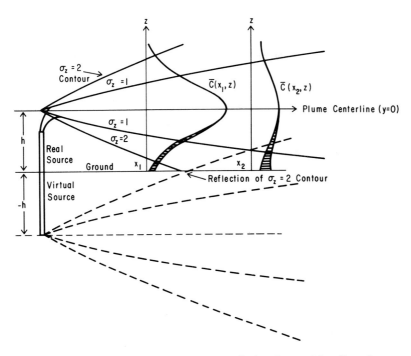

FIG. 1.6. Effect of "reflection" from the ground of a plume with a Gaussian concentration profile. Both real and virtual sources are shown, along with the contribution to the concentration due to reflection (shaded). The $\sigma_z = 1$ and $\sigma_z = 2$ contours are also shown, along with the $\sigma_z = 2$ contour due to reflection. h is the effective stack height. $y = 0$ is the plume centerline.

The two terms in z arise from consideration of "reflection" of the plume from the ground as shown in Fig. 1.6.

Graphs of σ_y and σ_z corresponding to various meteorological conditions are available for rapid estimation of atmospheric dispersion in Figs. 1.7a and 1.7b.

It is important to remind the reader that these estimates of diffusion are based on the assumption of no reaction or other removal or production mechanisms.

1.4.3 THE BOX MODEL

In some cases, it may be realistic to consider an extremely different sort of mixing. If material from a source is mixed thoroughly below an atmos-

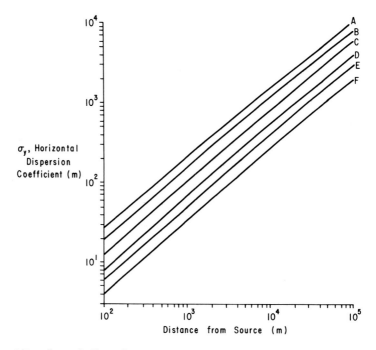

Fɪɢ. 1.7a.   Lateral dispersion σ_y as a function of downwind distance for various stability classes (Slade, 1968). (A) Extremely unstable. (B) Moderately unstable. (C) Slightly unstable. (D) Neutral. (E) Slightly stable. (F) Moderately stable.

pheric inversion (warm air overlying cool air) and there is no removal or production by other mechanisms, a "box" model may be used. In contrast to the Gaussian model, this model assumes that the concentration is constant in a plane oriented perpendicular to the mean wind as shown in Fig. 1.8.

This model is physically meaningful in those situations where the horizontal and vertical dispersion of the plume (or plumes) are limited by topographic features such as valley walls as well as by an inversion. Using the same nomenclature as before, we have

$$\bar{c}_{\text{box}}/Q' = 1/hy\bar{U} \tag{1.16}$$

Figure 1.9 illustrates the boxlike nature of this model if a source of strength Q' is mixed into a box of dimension $hy\bar{U}$. Here, h is the depth of the mixed layer, y is the width of the plume in the valley, $\bar{U}$ is the mean wind, Q' is the source strength, and $\bar{c}_{\text{box}}$ is the average concentration at the downwind end of the box.

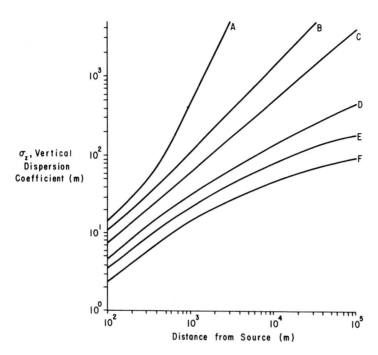

σ_z, Vertical Dispersion Coefficient (m)

Distance from Source (m)

FIG. 1.7b. Vertical dispersion σ_z as a function of downwind distance for various stability classes (Slade, 1968). Stability classes are the same as in Fig. 1.7a.

1.4.4 CLOUD PROCESSES

Owing to the existence of all three phases of water in the atmosphere, a complete family of processes occurs due to the interaction of trace materials with water in its different forms. From Fig. 1.1, we can list those processes that play an important (if complex) role in air chemistry:

1. Sorption of gases onto or into hydrometeors. (Hydrometeors are any sort of water particle much larger than molecular size in air.)

2. Deliquescence: That process which occurs when the vapor pressure of the saturated solution of a substance is less than the vapor pressure of water in the ambient atmosphere. Water vapor is collected until the substance is dissolved in an unsaturated solution in equilibrium with the environment.

3. Efflorescence: The loss of water by a solution droplet to form a solid particle. Efflorescence is thus the opposite of deliquescence. Changes in the physicochemical properties of a particle may occur when it undergoes a deliquescence–efflorescence cycle.

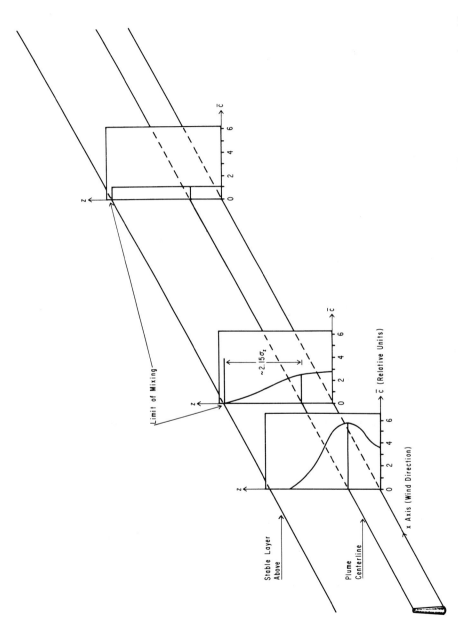

FIG. 1.8. Variation of concentration below a stable layer in vertical profile. The tendency toward a constant concentration with height is shown (Turner, 1969).

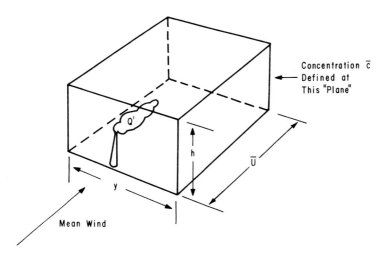

Concentration c̄
Defined at
This "Plane"

Fig. 1.9.   Box model for estimating concentration beneath a stable layer [see Eq. (1.16)]. (1) Mean wind; (2) the concentration c̄ is defined at this "plane."

4.   Raoult's equilibrium, which occurs when a solution droplet adjusts its composition so that its vapor pressure is equal to that of the environment. This occurs with deliquescent salts above their deliquescence point as well as with hygroscopic liquids such as H_2SO_4.

5.   Chemical reactions in the dissolved materials of solution droplets.

6.   Nucleation of cloud droplets when the air becomes sufficiently supersaturated to allow the particle to grow to large size.

7.   Evaporation of cloud drops to form high-humidity aerosols (such as during the "burnoff" of fog or stratus clouds).

8.   Capture of aerosol particles by falling rain or cloud drops or by the Brownian motion of particles. If the drop is condensing, the aerosol capture is enhanced. The opposite holds if the droplet is evaporating.

9.   Removal of materials from the air by precipitation (drizzle, rain, snow, etc.). This is often referred to as "rainout," and results both in the removal of material from air and the modification of composition of rain and snow water. Capture by falling raindrops, etc. leading to removal is referred to as "washout."

While these mechanisms are clearly important to the chemical behavior of air, they are sufficiently complex to be difficult to describe in a quantitative sense. Current research in the aerosol/cloud physics areas includes extensive studies of these processes, and the reader is referred to other sources for further study of them. Works on cloud physics (Mason, 1957;

Fletcher, 1962) and precipitation scavenging (Engelmann and Slinn, 1970) are useful in this regard.

1.4.5   A CAUTIONARY NOTE

Returning to the general equation for the rate of change of concentration, which is a form of the equation of continuity, Eq. (1.1),

$$dc/dt = Q + R_{prod} - R_{rem} - D - A - P$$

it is obvious that the diffusion models given here were based on the assumption that the equation could be reduced to a simple form that does not allow for any removal processes:

$$dc/dt = Q - D - A \tag{1.17}$$

Thus removal processes and *in situ* sources are *not* accounted for by normal diffusion models. Several classes of situations actually occur in the atmosphere where it is necessary to consider (even if only on a semiquantitative basis) such source and sink strengths. For simplicity, we can gather them under the terms in the equation.

(a)   *Production reactions.* Both gases and aerosols may be produced by reactions in the atmosphere. Materials so produced are sometimes referred to as "secondary" pollutants. Notable examples are O_3 (both in the upper atmosphere and in urban smog), NO_2, and sulfate aerosol (from SO_2).

(b)   *Removal reactions.* In some cases, but not all, reactions in the atmosphere produce one substance at the expense of another. For instance,

$$2SO_2 + O_2 + 2H_2O \rightarrow 2H_2SO_4$$

produces a sulfate aerosol while consuming SO_2. In contrast, O_3 production consumes only a minute fraction of the O_2. The rates of such reactions are variable, but can produce rates of change of concentration comparable to that due to diffusion, especially at some distance from the source where diffusion is not so effective. Unfortunately, it is not yet possible to accurately describe the reaction rates of atmospheric substances, so that the use of the continuity equation is seriously limited. Finally, it is necessary to remember that some gases react with and are removed by materials at the air-surface interface. Substances such as SO_2, O_3, and NO_2 react with plant surfaces, and SO_2 could be expected to be absorbed by either the wet surfaces after rain or dew or by bodies of water. Aerosols are removed by diffusion, settling, and impaction, once again when near the air-surface interface. They may also be reentrained by the wind, providing a secondary source of the material is involved.

(c) *Removal by hydrometeors and precipitation.* As pointed out above, it is not possible to quantitatively predict the effects of precipitation and cloud processes on removal of either gases or aerosol particles. Nonetheless, such removal does occur and must be considered.

One generalization can be made with regard to the effect of rain on aerosols. While it is popular to consider that rain "washes" or "cleanses" the air, and there is no doubt that rain water composition is often highly modified by falling through polluted air, the efficiency of removal of particles (and gases, for that matter) is extremely small. The clearing up of air following a rainstorm is more nearly the effect of a change of air mass or increased instability. The calculation of efficiency or removal by collision of droplets with aerosol particles involves calculation of Stokes' number, and is included in Chapter 9.

PROBLEMS

1. Show that $\bar{F} = \bar{U}\bar{c} + \overline{U'c'}$. What are the necessary conditions for the statement that $\bar{F} = \bar{U}\bar{c} + \overline{U'c'}$?
2. Apply the box model of plume diffusion to a 500 ton/day source of SO_2 located at the windward end of a 10 km wide valley under a 500-m inversion. Calculate the average concentration of SO_2 in ppm and the concentration of SO_4^{2-} in $\mu g/m^3$, assuming complete oxidation of the SO_2. Assume a wind speed of (a) 5 km/hr; (b) 35 km/hr.
3. Derive Eq. (1.16). State all assumptions. At what points in the box shown in Fig. 1.9 would you expect this approximation to be most valid?

REFERENCES

Berkner, L. V., and Marshall, L. C. (1964). *In* "The Origin and Evolution of the Atmospheres and Oceans" (P. J. Brancazio and A. G. W. Cameron, eds.). Wiley, New York.

Broecker, W. S. (1970). *Science* **168,** 1537.

Engelmann, R. J., and Slinn, W. G. N. (1970). "Precipitation Scavenging (1970)." U.S. At. Energy Comm., Oak Ridge, Tennessee.

Fleagle, R. G., and Businger, J. A. (1963). "An Introduction to Atmospheric Physics," Academic Press, New York.

Fletcher, N. H. (1962). "The Physics of Rainclouds." Cambridge Univ. Press, London New York.

Friend, J. P., and Charlson, R. J. (1969). *Environ. Sci. Technol.* **3,** 1181.

Healy, T. V., McKay, H. A. C., Pilbeam, A., and Scargill, D. (1970). *J. Geophys. Res.* **75,** 2317.

LaHue, M. D., Pate, J. B., and Lodge, Jr., J. P. (1970). *J. Geophys. Res.* **75,** 2922.

Machta, L., and Hughes, E. (1970). *Science* **168,** 1582.

Mason, B. J. (1957). "The Physics of Clouds." Oxford Univ. Press (Clarendon), London and New York.

Rubey, W. W. (1951). *Bull. Geol. Soc. Amer.* **62,** 1111.

Slade, D. H. (1968). "Meteorology and Atomic Energy-1968." U.S. At. Energy Comm., Oak Ridge, Tennessee.

SMIC (1971). Inadvertant climate modification. Rep. of the Study of Man's Impact on Climate. MIT Press, Cambridge, Massachusetts.

Swinnerton, J. W., Linnenbom, V. J., and Cheek, C. H. (1969). *Environ. Sci. Technol.* **3,** 836.

Turner, D. B. (1969). Workbook of atmospheric dispersion estimates. Publ. No. 999-AP-26. Public Health Service, Cincinnati, Ohio.

U.S. Committee on Extension to the Standard Atmosphere (1962). "U.S. Standard Atmosphere, 1962." U.S. Gov't. Printing Office, Washington, D.C.

Van Valen, L. (1971). *Science* **171,** 439.

Weast, R. C., ed. (1968). "Handbook of Chemistry and Physics," 49th ed. Chem. Rubber Co., Cleveland, Ohio.

CHAPTER 2

SUMMARY OF CHEMICAL PRINCIPLES

The purpose of this chapter is to review the chemical principles which will be required for an understanding of some of the material in the rest of the book. There is no attempt to summarize all of the chemistry relevant to the study of air chemistry; all of the material considered in this chapter will be used later in the book.

The person with some background in chemistry may treat this chapter as a review. The treatment given to some of the topics in this chapter may be too concise for those without the background of a college chemistry course. A variety of references are included at the end of the chapter. There is also a selection of self-study materials available in the field of general chemistry.

The first four sections of the chapter are concerned with the properties of systems at equilibrium. These principles reappear in considerations of the analysis of compounds in the atmosphere and in the relationships between the concentrations present in the atmosphere and in other phases. Extinction of radiation is a principle which reappears in photometric analytical methods, photochemical processes, and consideration of atmospheric extinction due to gases and aerosols. The chemical kinetics examined here must be combined with the atmospheric processes considered in Chapter 1 in accounting for rates of change of concentrations in the atmosphere.

2.1 Properties of Gases

To a high degree of approximation, at atmospheric conditions, the state of a substance in the gas phase may be characterized by the ideal gas equation:

$$PV = nRT \tag{2.1}$$

The units in which the ideal gas constant R is expressed will depend upon the units chosen for the variables P, V, n, and T. The temperature T is nearly always expressed in degrees Kelvin, K. Although observed temperatures are often reported in °C or °F, the T which appears in mathematical expressions in this book will always be in degrees Kelvin. The quantity n is measured in moles (we shall use the term "moles" for "gram-moles") and V is often expressed in liters, although it is sometimes more conveniently measured in cubic meters (m³) or cubic centimeters (cm³). The pressure may be expressed in a wide variety of units: The atmosphere (1 atm = 1,013,250 dyn/cm²) is a common unit, as is the Torr (760 Torr = 1 atm). Aside from small corrections due to thermal expansion terms, the pressure in Torr corresponds to the height of a column of mercury, measured in millimeters, which the gas will support. The millibar (mb), often used by meteorologists and other atmospheric scientists, is defined as 1000 dyn/cm². Some values for R in various units are given in Table 2.1.

Dalton's law states that, in a mixture of ideal gases, each gas exerts a pressure which depends only on the state of that particular gas:

$$P_A V = n_A RT$$

This pressure P_A is frequently referred to as a partial pressure. The total pressure of all the gases in a mixture is the sum of the partial pressures:

$$P_{tot} = \sum P_A$$

Therefore, in a mixture of ideal gases,

$$P_{tot} V = n_{tot} RT$$

TABLE 2.1

IDEAL GAS CONSTANT

$$R = 8.3143 \times 10^7 \text{ ergs mole}^{-1} \text{ K}^{-1}$$
$$= 1.9872 \text{ cal mole}^{-1} \text{ K}^{-1}$$
$$= 82.056 \text{ cm}^3 \text{ atm mole}^{-1} \text{ K}^{-1}$$

Other variables which are considered in connection with the ideal gas equation are the molar concentration and the mass concentration, or density. The molar concentration is simply n/V. Density is usually represented by the symbol ρ $(= m/V$, where m is the mass of the gas being considered). The ideal gas equation in terms of density may be written

$$P_A = \rho RT/M_A \tag{2.2}$$

where M_A is the molecular weight of the gas.

In some studies, quantities of matter are expressed in terms of molecules/cm³, N_A, rather than moles, n_A. These quantities are related by Avogadro's constant, N_0:

$$N_A = N_0(n_A/V)$$

Avogadro's number is the number of molecules in one mole of a substance: $N_0 = 6.02252 \times 10^{23}$ molecules/mole. The units of N_0 are not often used, but are convenient for dimensional analysis.

2.2 Properties of Solutions

Raoult's law describes the pressure of the vapor of a substance in equilibrium with a solution of the same substance:

$$P_A = X_{lA}P_A^\circ \tag{2.3}$$

X_{lA} is the mole fraction of the substance in solution and P_A° is the vapor pressure of the pure substance A. The quantity X_{lA} is defined as the number of moles of A in solution divided by the total number of moles in solution, $X_{lA} = n_A/n_{tot}$. Raoult's law is most often applied in those cases where the liquid phase is mostly substance A. For ideal solutions, Raoult's law may be applied over the complete range of composition; for most real solutions this approximation becomes worse as X_{lA} becomes small. Values for P_A° may be found in the reference sources listed in the bibliography; Table 2.2 lists values for water.

Henry's law is similar in mathematical form to Raoult's law:

$$P_A = k_H X_{lA} \tag{2.4}$$

P_A and X_{lA} retain their previous meanings. The Henry's law constant k_H depends on temperature, substance A, and the solvent in which A is dissolved. It should be pointed out that the dimensions of k_H will depend on the units in which P_A is expressed and the concentration units for A in

TABLE 2.2

VAPOR PRESSURE OF WATER[a]

T (°C)	0	5	10	15
P (Torr)	4.579	6.543	9.209	12.788

T (°C)	20	25	30	35
P (Torr)	17.535	23.756	31.824	42.175

[a] Weast, R. C., ed. (1970). "Handbook of Chemistry and Physics," Chem. Rubber Co., Cleveland, Ohio.

the liquid phase. Some references may express P_A in Torr and composition in molarity or grams per liter. Some values for Henry's law constants are collected in Table 2.3, in which the pressure is expressed in atmospheres and the composition is expressed as a mole fraction.

2.3 Properties of Small Droplets and Particles

The pressure of a vapor in equilibrium with a liquid or solid also depends on the radius of curvature of the surface of the condensed phase. Stated without derivation, the relationship between the pressure of the small droplet, P_A, and the pressure of the bulk phase, $P_A{}^\circ$, is given by the expression

$$RT \ln(P_A/P_A{}^\circ) = 2\bar{V}\sigma/r \tag{2.5}$$

$\bar{V}$ is the volume per mole of the liquid, or condensed phase, σ is the surface tension, and r is the radius of the small particle. The increase of vapor pressure as the particle radius decreases is known as the Kelvin effect.

TABLE 2.3

HENRY'S LAW CONSTANTS (IN ATM) AT 25°C OF SELECTED GASES IN H_2O[a]

CO_2	1.64×10^3	CH_4	4.1×10^4
CO	5.8×10^4	NO	2.9×10^4
SO_2	39.7	N_2O	2.3×10^3
NH_3	3.0	H_2S	5.4×10^4

[a] Washburn, E. W., ed. (1926). "International Critical Tables." McGraw-Hill, New York.

A thermodynamically analogous situation exists when small particles, such as atmospheric aerosol, are trapped in liquid water. The curvature of the surface of the particle affects the solubility in the same sense as the Kelvin effect. That is, small particles are more soluble (have a higher activity) than bulk material, which has a very large radius of curvature. This increased solubility is potentially important in cloud and precipitation processes, in deliquescence phenomena, and in sampling where particles are trapped in liquid reagents.

2.4 Chemical Thermodynamics

When substances undergo chemical reactions, the concentrations of the reacting species at equilibrium may be described in terms of their thermodynamic properties. The equilibrium constant for the chemical reaction $aA + bB \rightleftarrows cC$ is defined as

$$K_{eq} = (a_C)^c / (a_A)^a (a_B)^b \tag{2.6}$$

where a_A is the activity of reactant A, and so on. It is important to specify the state of each reactant and product in the reaction since the expression for the activity will depend on whether that substance is present in the gas, solid, or liquid state.

If the substance A is in the gas phase and if A behaves ideally, the activity of A may be replaced by the partial pressure P_A measured in atmospheres. For cases in which A occurs as the solute in a *dilute* solution, the activity may be replaced by the concentration in moles/liter, usually written [A], if A forms ideal solutions. When reactant A is the solvent in a liquid phase, the activity may be replaced by the mole fraction X_A for solutions which obey Raoult's law. The activity of A is set equal to unity when A appears as a pure liquid or a pure solid. (This is the approximation made for the activity of water in most ionic equilibrium expressions.)

Although significant departures from ideality may be observed in many systems of interest, for the present purposes, we shall be concerned only with ideal cases in which the above-mentioned substitutions for activity may be made.

While K_{eq} is tabulated for many common reactions, it is more often derived from thermodynamic variables. From considerations of the second law of thermodynamics [see, for example, Moore (1972), and Klotz (1964)], we may show that

$$\ln K_{eq} = -\Delta G_T^\circ / RT \tag{2.7}$$

where

$$\Delta G_T° = \Delta H_T° - T \, \Delta S_T° \tag{2.8}$$

G, H, and S are the thermodynamic functions of state: Gibbs free energy, enthalpy, and entropy, respectively. Quantities from which these thermodynamic variables may be derived are tabulated in many tables. The symbol Δ refers to the change

$$\text{Reactants} \rightarrow \text{Products}$$

The degree sign means that the change in the thermodynamic function is to be evaluated when reactants and products are in their respective standard states. The subscript T refers to the temperature at which the reaction is being considered. The free energy of formation of a substance, $\Delta G°_{fT}(A)$, is defined as the change in G when one mole of A is formed from its constituent elements, each reactant being in its standard state. If the free energies of formation $\Delta G°_{fT}(A)$ are available for each of the reactants, then

$$\Delta G° = c \, \Delta G_{fT}(C) - a \, \Delta G_{fT}(A) - b \, G°_{fT}(B) \tag{2.9a}$$

and K_{eq} may be evaluated directly. We can also evaluate K_{eq} if the enthalpies of formation $\Delta H°_{fT}$, and standard entropies $S_T°$ are available. The terms $\Delta H_T°$ and $\Delta S_T°$ are defined by equations analogous to Eq. (2.8a) in terms of $\Delta H°_{fT}(A)$ and $S_T°(A)$:

$$\Delta H_T° = c \, \Delta H°_{fT}(C) - a \, \Delta H°_{fT}(A) - b \, \Delta H°_{fT}(B) \tag{2.9b}$$

$$\Delta S_T° = c \, S_T°(C) - a \, S_T°(A) - b \, S_T°(B) \tag{2.9c}$$

Equation (2.9a) may be understood in terms of the following diagram:

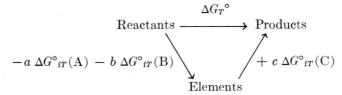

The fact that G is a function of state means that the change in G, ΔG, in going from one state to another is independent of the path taken. In this example, ΔG will be the same if we proceed from reactants to elements and then to products as it will be if we go directly from reactants to products.

In many tables which evaluate thermodynamic functions at higher than ambient temperatures, the quantities $(G_T - H_{298})/T$ and $(H_T - H_{298})/T$ are tabulated instead of ΔG_{fT} and $\Delta G°_{fT}$ and $\Delta H°_{fT}$. The former

TABLE 2.4

FREE ENERGIES OF FORMATION OF
COMMON ATMOSPHERIC COMPOUNDS[a]

	$\Delta G°_{f298}$		$\Delta G°_{f298}$
CO_2	−94.265	NO	+20.769
CO	−32.783	NO_2	+12.403
SO_2	−71.741	N_2O_4	+23.485
SO_3	−88.545	H_2O	−54.636
NH_3	− 3.966	O_3	+38.997
N_2O	+24.780		

[a] In kcal mole⁻¹. Stull, D. R., and
Prophet, H. (1971). "JANAF Ther-
mochemical Tables," 2nd ed. Nat.
Bur. Stand. Washington, D.C.

quantities change more slowly with temperature and may be determined
for intermediate temperatures by interpolation of points given in the tables.
$\Delta G_T°$ may be determined from the relationship

$$\frac{\Delta G_T°}{T} = \frac{\Delta(G_T° - H°_{298})}{T} + \frac{\Delta H°_{298}}{T}$$

$\Delta H°_{298}$ may be determined using Eq. (2.9b) and the heats of formation,
which are commonly given for 298 K. The standard free energy of formation and standard heats of formation of any element in its standard state
is zero. The free energies of formation of some of the gases commonly encountered in the atmosphere are given in Table 2.4. Additional values of
thermodynamic variables may be found in Rossini *et al.* (1952) and Stull
et al. (1969).

2.5 Extinction of Radiation

The loss of intensity in a beam of radiation passing through a substance
may be described by the Beer–Lambert–Bouguer law:

$$-dI_\lambda/dz = c\alpha_\lambda I_\lambda \tag{2.10}$$

In this expression, I_λ is the intensity of radiation at wavelength λ; c is the
concentration of the substance causing the attenuation of light in moles
per unit volume; z is the distance in the sample through which the light
passes; and α_λ is the natural molar extinction coefficient. The integrated

form of this equation is

$$I = I_0 \exp(-\alpha_\lambda c z) \tag{2.11}$$

where I_0 is the radiation incident on the sample. It can be seen from this expression that the units of α_λ will depend on the units chosen for c and z. It should be noted at this point that chemists in dealing with absorbing systems usually express this law in the form

$$I = I_0 10^{-\epsilon_\lambda c z}$$

where ϵ_λ ($= \alpha_\lambda/2.303$) is known as the absorptivity. Absorbance, defined as $\log(I_0/I_\lambda)$, is directly proportional to concentration and is usually the quantity indicated by instruments designed to determine concentration from a measurement of light absorption. The light extinction may also be expressed in terms of transmittance, I/I_0, often given as a percentage.

Returning to Eq. (2.10), it should be pointed out that if there are a number of species present in the sample and if we recognize that extinction can result from absorption and from scattering, then $c\alpha_\lambda$ must be replaced by a sum which includes contributions from all species:

$$
\begin{aligned}
c\alpha_\lambda &= b_{\text{abs}} + b_{\text{scat}} \\
&= \sum_i c_i(\alpha_{\text{a}\lambda i} + \alpha_{\text{s}\lambda i}) \tag{2.12}
\end{aligned}
$$

The summation extends over all species. The terms $c_i\alpha_{\text{a}\lambda i}$ and $c_i\alpha_{\text{s}\lambda i}$ are the absorption coefficient and scattering coefficients, respectively; b_{abs} and b_{scat}, which represent, respectively, absorption and scattering per unit length, are also mentioned here because these quantities are often the ones determined from measurements of extinction in the atmosphere.

2.6 Chemical Kinetics

Chemical kinetics is concerned with the time dependence of concentration. The time dependence is often expressed in terms of a differential equation which relates the rate of change of concentration with respect to time, dc/dt, to a function of the concentrations of reacting species. Although the overall reactions of many substances in the atmosphere are complex and involve a large number of reactants, one usually assumes that these complex reactions may be represented by the simultaneous occurrence of a number of elementary reactions. Thus, while the oxidation of nitric oxide to nitrogen dioxide in urban atmospheres is a complex reaction which may involve the participation of sunlight, organic molecules, ozone, water vapor, and other trace constituents, in an addition to a host of highly reactive intermediate substances, we assume that this reaction may be understood in terms of elementary reactions such as the reaction between an oxygen molecule and two nitric oxide molecules.

$$O_2 + 2NO \rightarrow 2NO_2 \qquad\qquad I$$

(Current evidence suggests that while this particular reaction may play a minor role in the atmospheric oxidation of nitric oxide, it is probably not of primary importance. Although it is also a complex reaction, we will consider it to be a simple reaction for purposes of illustration.) The rate of this reaction depends only on the concentrations of the reactants NO and O_2 and may be written

$$-d(O_2)/dt = k_f(O_2)\,(NO)^2$$

k_f is the *rate constant* for this reaction. The minus sign is included so that k_f will be positive. The dimensions of k_f will depend on the concentration and time units. In many laboratory studies, k_f for this reaction would be expressed in $(mole/liter)^{-2}\,sec^{-1}$. In air pollution studies, it might be more convenient to express k_f in $ppm^{-2}\,hr^{-1}$. In the rate expression for an elementary reaction, the concentration of each reactant is raised to the power of the number of molecules of that reactant entering the reaction. The order of the reaction is the sum of all exponents in the rate expression, or the total number of reactant molecules. This is a third-order reaction, second order with respect to nitric oxide and first order with respect to oxygen. The rate of this reaction may also be described in terms of the rate of change of (NO) or (NO$_2$):

$$d(NO)/dt = 2d(O_2)/dt$$

$$d(NO_2)/dt = -2d(O_2)/dt$$

The temperature dependence of a rate constant is often expressed by the Arrhenius equation:

$$k = A \exp\,(-E_a/RT) \qquad\qquad (2.13)$$

This is an empirical relationship which satisfactorily accounts for the temperature dependence, over small temperature ranges, of a large number of rate constants. Here, A is the preexponential factor, or frequency factor, and E_a is the activation energy. Many sources tabulate $\log(A)$ and express E_a in kcal/mole.

The reverse reaction of I is

$$2NO_2 \rightarrow 2NO + O_2 \qquad\qquad II$$

for which

$$d(NO)/dt = 2k_r(NO_2)^2$$

If we allow the forward and reverse reactions to proceed simultaneously, the rate of change of (NO) will be

$$d(NO)/dt = -2k_f(O_2)\ (NO)^2 + 2k_r(NO_2)^2 \tag{2.14}$$

The first term accounts for the decrease of (NO) due to reaction I, the second term accounts for the formation of NO in reaction II. If there are other reactions being considered which involve NO, a rate term for each of these reactions will be added to the right-hand side of (2.14), with a positive or negative sign according to whether NO is formed or decomposed in the particular reaction being considered.

In the rather special case in which I and II are the only reactions involving NO being contemplated, if (NO) is constant with respect to time, $d(NO)/dt = 0$ and

$$k_f/k_r = (NO_2)^2/(NO)^2(O_2) = K_{eq}$$

This may also be recognized as an equilibrium constant expression. The reader can show that for a more general case with the simultaneous reactions

$$aA + bB \rightarrow cC \qquad \text{and} \qquad cC \rightarrow aA + bB$$

an expression similar to (2.6) is obtained if $dA/dt = 0$.

The condition that the time derivative of concentration of a substance vanish is known as the steady-state condition under the following circumstances. Suppose that the reaction $A \rightarrow C$ is represented by the two reactions

$$A \rightarrow B \qquad k_1 \qquad \text{and} \qquad B \rightarrow C \qquad k_2$$

The condition $dB/dt = 0$ is known as a steady-state condition and may be applied in some cases when the total system is not in equilibrium. If this is the case for this system, then $dC/dt = k_1A$.

Although it is not the purpose of the present work to discuss detailed aspects of reaction rates, photochemical processes are of sufficient importance to the study of this subject to justify a few comments. The rate of a primary photochemical process, such as the decomposition of NO_2,

$$NO_2 + h\nu \rightarrow NO + O$$

may be written

$$-d(NO_2)/dt = \phi I_a \tag{2.15}$$

where I_a is the rate of photon absorption per unit volume by NO_2 in the wavelength region of interest and ϕ is the primary quantum efficiency, the fraction of photon absorptions that lead to the production of $NO + O$.[*] Using the Beer–Lambert–Bouguer law, the probability that a photon will be absorbed in distance z is $\alpha_\lambda(NO_2)z$. The number of absorptions in a volume of cross section s and length z per unit time is then $\alpha_\lambda(NO_2)F_0sz$, where F_0 is the number of moles of photons (einsteins) per unit area per unit time incident on the sample of NO_2. It is proportional to the light intensity integrated over all angles as seen by the sample of gas, $F_0 \propto \int I_0\,d\omega$. From this, it may be seen that $I_a = \alpha_\lambda(NO_2)F_0$ and

$$-d(NO_2)/dt = \phi\alpha_\lambda F_0(NO_2) \qquad (2.16)$$

The term $\phi\alpha_\lambda F_0$ will have the dimensions of a first-order rate constant, sec^{-1}. This expression is valid only for monochromatic radiation and may be applied in the present form only to laboratory studies. For atmospheric processes, it must be recognized that ϕ, α_λ, and F_0 all depend on the wavelength and that the product of these terms must be integrated over the entire spectrum. This task is simplified somewhat because ϕ usually approaches zero in the long-wavelength region of the spectrum and F_0 goes to zero in the short-wavelength region. In real life, F_0 will not only contain a contribution due to direct solar radiation, attenuated by the atmosphere, but will also include contributions from radiation scattered and reflected from the atmosphere and from the earth. Leighton (1961) has discussed some of the contributions to F_0 in greater detail than we can present here. Calvert and Pitts (1966) describe the photochemistry of a number of compounds and the measurement of F_0. Other than the wavelength dependence, α_λ and ϕ are independent of external factors and are functions only of the absorbing molecule.

In the atmosphere, photolytic reactions and the spontaneous decomposition of reactive intermediates are generally first-order reactions. Second-order reactions of the type $A + C \rightarrow C + D$ are expected for reacting molecules of all sizes. Third-order reactions are somewhat rare, but they play an important role in the combination of small molecules. When an oxygen atom and an oxygen molecule combine to form ozone, $O + O_2 \rightarrow O_3$, the large amount of energy released when the new bond is formed often cannot be redistributed in the product molecule before it causes the mole-

---

[*] It should be noted that in discussions of the Beer–Lambert–Bouguer law, intensity usually has the dimension energy cm^{-2} sec^{-1} steradian^{-1}. In the present discussion, I_a is not an intensity, but rather the number of moles of photons absorbed per unit volume per unit time by NO_2.

cule to break apart again. The presence of a third body M may allow some of this energy to be carried off, thus stabilizing the product molecule. The reaction is then of third order and is written

$$M + O + O_2 \rightarrow O_3 + M$$

Larger product molecules can sometimes be stabilized without the presence of a third body.

PROBLEMS

1. Calculate a in the expression

$$c' = aMc$$

at 25°C, where c' is the concentration in $\mu g/m^3$, c is the mole fraction in ppm, and M is the molecular weight in grams of the gas.

2. Imagine a closed system containing only CO_2 at 1 atm and H_2O at 25°C. Calculate the mole fraction of CO_2 dissolved in water and the partial pressure of water in the vapor phase, given $k_H(CO_2) = 1.6 \times 10^3$ atm, $P°(H_2O) = 23.8$ Torr.

3. Calculate the partial pressure of N_2O_4 in equilibrium with NO_2 [$2NO_2(g) = N_2O_4(g)$; $\Delta G°_{f298}(NO_2) = 12.39$ kcal/mole; $\Delta G°_{f298}(N_2O_4) = 23.49$ kcal/mole] (a) when $P_{NO_2} = 0.5$ atm; (b) when $P_{NO_2} = 1 \times 10^{-6}$ atm.

4. Calculate the partial pressure of NO in equilibrium ($N_2 + O_2 = 2NO$) with a standard atmosphere of N_2 and O_2 (78% N_2, 21% O_2) (a) at 298 K, (b) at 1500 K, given the values in the accompanying table:

	$-(G_T° - H°_{298})/T$ (cal mole^{-1} K^{-1})		$\Delta H°_{f298}$
	$T = 298$ K	$T = 1500$ K	(kcal mole^{-1})
N_2	45.770	51.665	0
O_2	49.004	55.185	0
NO	50.347	56.442	21.652

5. Johnston and Crosby (1951) report the following parameters for the reaction $O_3 + NO \rightarrow NO_2 + O_2$: $\log(A) = 11.90$ (A in mole^{-1} cm^3 sec^{-1}, $E_a = 2.5$ kcal/mole. Calculate the rate constant for this reaction at 25°C in pphm^{-1} hr^{-1}.

6. The concentration of SO_2 in unpolluted continental air is frequently of the order of 30 $\mu g/m^3$. Express this concentration in: (a) ppm; (b) moles/liter; (c) molecules/cm^3; (d) $\mu g/m^3$ of $(NH_4)_2SO_4$.

7. SO_2 may be determined by a colorimetric method in which the concentration of SO_2 is related to the absorbance of the solution used to absorb the SO_2 from the atmosphere. In a calibration of this method, 100 μg of SO_2 gave a solution with a transmittance of 45.5%. The absorption of the SO_2 in 300 liters of air at 30°C in the same volume of absorbing solution yielded a transmittance of 56%. Calculate the concentration of SO_2 in the sample of air in: (a) ppm and (b) $\mu g/m^3$ at 25°C.

REFERENCES

A general introduction to most topics in this chapter may be obtained from any of a number of college freshman level chemistry texts. More extensive treatments of the topics in this chapter are the following.

Benson, S. W. (1960). "The Foundations of Chemical Kinetics." McGraw-Hill, New York.

Calvert, J. G., and Pitts, J. N. (1966). "Photochemistry." Wiley, New York.

Frost, A. A., and Pearson, R. G. (1961). "Kinetics and Mechanism," 2nd ed. Wiley, New York.

Johnston, H. S., and Crosby, H. J. (1951). *J. Chem. Phys.* **19,** 799.

Klotz, I. M. (1964). "Introduction to Chemical Thermodynamics." Benjamin, New York.

Leighton, P. A. (1961). "Photochemistry of Air Pollution." Academic Press, New York.

Moore, W. J. (1972). "Physical Chemistry," 4th ed. Prentice-Hall, Englewood Cliffs, New Jersey.

REFERENCES FOR TABLES

Rossini, F. D., Wagman, D. D., Evans, W. H., Levine, S., and Jaffe, I. (1952). Selected values of chemical thermodynamic properties. *Nat. Bur. Stand. (U.S.) Circ.* **No. 500.**

Stull, D. R., and Prophet, H. (1971). "JANAF Thermochemical Tables," 2nd ed. Nat. Bur. Stand., Washington, D.C.

Stull, D. R., Westrum, E. F., and Sinke, G. C. (1969). "The Chemical Thermodynamics of Organic Compounds." Wiley, New York.

Washburn, E. W., ed. (1926). "International Critical Tables." McGraw-Hill, New York.

Weast, R. C., ed. (1970). "Handbook of Chemistry and Physics," 51st ed. Chem. Rubber Co., Cleveland, Ohio.

CHAPTER 3

SAMPLING AND COLLECTION

In the present chapter, we shall consider methods used to collect samples. The consideration of the placing of sampling points is beyond the scope of this book. Suffice to say that careful attention must be given to meteorological and topographic factors in addition to the nature of sources and receptors when choosing measurement sites. In addition to measurements of chemical factors, such variables as wind velocity, temperature, barometric pressure, and relative humidity should also be available. The factors which have been considered in the selection of monitoring sites are described by Leavitt *et al.* (1957), Stalker *et al.* (1962), and Craw (1970).

3.1 Generalized Methods of Analysis

Sampling and collection methods depend on the sort of measurement being made. Measurement types may be broadly divided into two groups: remote measurements and measurements requiring physical collection or handling of the sample.

Remote methods are generally most desirable in that measurements may be made over a wide area from a single location and they may be made in areas where sample collection is difficult or expensive, such as the top of a smokestack or the middle of an ocean. One of the earliest remote techniques was the Ringelmann method (U.S. Bur. Mines, 1967) for estimating the blackness of a smoke plume. Recent developments include the use of LIDAR (Barrett and Ben-Dov, 1967) for the study of aerosols and correlation spectrophotometry (Moffat and Millan, 1971) for the remote detection of gaseous constituents. While the advantages of remote measurements are obvious, more extensive use will require instruments of greater sensitivity than are now available.

Methods which require some manipulation of the sample may also be divided into two categories: (1) those that make a direct physical measurement of some property of the sample *in situ*; (2) those that first require an extraction step before making a physical or chemical measurement. The practical distinction between these methods is that no knowledge of the sample volume or sampling rate is required for the first type of measurement; for the second type, one must have information on the sampling rate or the total volume of air sampled in order to infer something about the atmospheric concentration of the species measured.

Typical of the methods that do not require knowledge of the sampling rate are nondispersive infrared instruments commonly used for the measurement of CO and CO_2; dispersive infrared spectroscopy, which has been used to study atmospheric samples in multiple reflection cells; and the integrating nephelometer (Charlson *et al.*, 1969) which measures the scattering coefficient in the atmosphere. Further details of instruments mentioned here are discussed in Chapters 5, 8, and 9. Although care must be taken in the use of these methods to ensure that the sample reaching the instrument is representative of the atmosphere, the measurements obtained are independent of the rate at which the atmosphere is sampled and therefore the rate does not have to be carefully controlled.

Typical of methods which do depend on the sampling rate is that which is commonly used in the determination of NO_2 (U.S. Dept. of Health, Education, and Welfare, 1965). In this method, NO_2 is extracted from the atmosphere by bubbling air through a reagent which traps the NO_2 in the form of an intensely colored azo dye. The concentration of the azo dye is then determined by measuring the absorbance of the solution. It is clear that the atmospheric concentration of NO_2 inferred from such a measurement will depend on the efficiencies of the chemical reactions and removal of NO_2 from the atmosphere; most importantly, it will require knowledge of the flow rate of air and reagent through the system.

3.2 Sampling Methods

After removal of the sample from the atmosphere, the analysis may be completed on the site or in the laboratory. Measurement on the site requires only a method for moving the atmospheric gas through the instrument. The essential components of this type of sampling are shown in Fig. 3.1.

If aerosol properties are being studied, one must keep the flow velocity and design of interior surfaces in mind to avoid loss of the sample by impaction. If gases are being studied, the inlet may contain components to eliminate interfering gases. In this case, care must be taken to ensure that these components do not deteriorate. The inlet system must be chemically and physically inert with respect to the substance of interest. Chemical inertness means mostly that the surfaces of the inlet system must not react with the sample. A change in the light intensity in the instrument may affect the composition of the sample if there are photochemically active compounds present (Butcher and Ruff, 1971). In some cases, changes in temperature or pressure may alter the composition of the sample. It is also necessary to protect the inlet from precipitation and to prevent the formation of fog within the instrument or inlet due to temperature changes. This is of particular importance in air-conditioned (cooled) locations. Yamada and Charlson (1969) suggest a large diameter (10–15 cm) pipe for sampling to minimize losses of reactive substances to the walls of the inlet system.

3.2.1 WHOLE AIR SAMPLES

"Grab" samples are those which are collected at one site and removed for study elsewhere. Glass, metal, or plastic film containers may be used. The glass container shown in Fig. 3.2 may be connected to a vacuum system or other instrument in the laboratory. The sample is collected by opening a previously evacuated container and allowing the air to rush in or by opening both stopcocks and pumping air through the flask.

Since the sample will generally be in contact with the container walls for a few hours, it is essential that the container material be inert. The flow-through method is generally preferred to the evacuated bulb method

Fig. 3.1. Generalized analytical system.

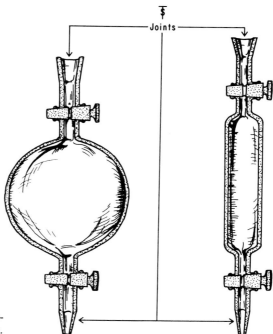

FIG. 3.2.   Flow-through collection flasks for whole air samples.

because it is simpler, safer, and because the walls may be conditioned to the atmosphere before collecting the sample.

Because of the expense and mass of large, rigid gas containers, large plastic bags have also been used to collect grab samples. The properties of a variety of materials from which these bags may be constructed have been summarized by Altshuller *et al.* (1962, 1970). These bags may be filled by pumping air through the bag, by expanding the bag by hand, or by using a rigid outer container as shown in Fig. 3.3. In this latter method, the bag may be emptied by increasing the pressure inside the box, and filled by reversing the vacuum blower so that the pressure inside the box is reduced. The choice of method used in collecting grab samples will often be determined by the volume of air required for analysis. Rasmussen (1970) has used cryogenic methods for collecting whole air samples. The stainless steel trap and heat exchanger shown in Fig. 3.4 are maintained at the temperature of boiling liquid air. All components in a given volume of the atmosphere are collected without the aid of pumps. The sampling rate may be controlled with a critical orifice. Although a whole air sample is collected, the analysis is usually performed on a few constituents of interest by allowing the volatile gases to boil off at controlled temperatures.

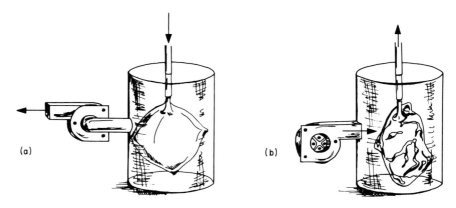

Fig. 3.3.   Arrangements for (a) filling and (b) emptying nonrigid sample bags.

3.2.2   EXTRACTION OF THE SAMPLE

In many analyses, it is convenient to obtain a hundredfold or more concentration of the trace constituents by getting rid of the oxygen, nitrogen, and, perhaps, carbon dioxide and water. The simplest concentration methods do not affect the chemical state of the sample in an irreversible fashion.

Most gaseous constituents may be concentrated by simply passing the air sample through a cold trap maintained at $-196°C$ by liquid nitrogen. At this temperature, most substances are solids. Nitrogen and oxygen may be eliminated with a vacuum pump at $-196°C$ or simply boiled off by raising the temperature 10–20°C. Carbon dioxide may be pumped off without losing most organic compounds by raising the temperature to $-100°C$. A wide range of low-temperature baths may be made up for the physical separation of trace constituents (Jolly, 1960). Care must be taken in the collection step of the cold trap method to avoid the formation of mists which may blow through the trap. Glass beads, glass wool, and other inert packing materials have been used to increase the efficiency of this method. Traps similar to the one shown in Fig. 3.5 and used by Shepherd *et al.* (1951) have been used.

Water and carbon dioxide may be eliminated by the use of selective reagents rather than distillation; however, one must be certain that these reagents do not interfere with the compounds of interest. Quiram and Biller (1958) used Ascarite and phosphorous pentoxide to eliminate carbon dioxide and water, respectively, and recovered more than 95% of the hydrocarbons in a synthetic sample with a cold trap. The efficiency of this method does depend on the length of time the sample is exposed to the cold surfaces. Hoshino *et al.* (1964) increased the ease of operation of this method by recirculating the air through the cold trap with a magnetic pump.

Chemical adsorption has the advantage of being somewhat more selective than simple cold traps. In this method, the trace constituents are adsorbed on a solid at a relatively low temperature and then desorbed at higher temperatures and low pressures. Quiram *et al.* (1954) used silica gel at $-73°C$ to adsorb hydrocarbons. Carbon dioxide was not adsorbed at this temperature and water was first eliminated with a drying tube; desorption was accomplished by heating gently with a Bunsen burner. The efficiency of this method is 95% for hydrocarbons containing three or more carbon atoms, but drops to 60% for ethane. Adams *et al.* (1960) used silica gel at $-78°C$ to adsorb sulfur compounds after first eliminating water. Desorption was carried out at $100–130°C$ for 1 hr and efficiencies were greater than 90% for CH_3SH, H_2S, and SO_2. For any new application of this method, it is important to make sure that the adsorption–desorption process does not affect the compound of interest. The present discussion is intended only to indicate a few of the possibilities of adsorption. The

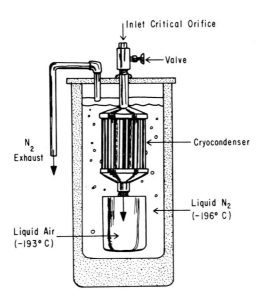

FIG. 3.4. Cold trap for the collection of whole air samples. (Redrawn with the permission of Dr. R. A. Rasmussen.)

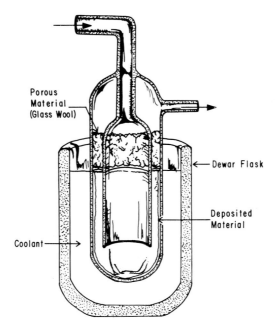

Fig. 3.5.   Flow-through cold trap. Porous material is used to reduce entrainment of mists.

interested reader is directed to one of the bibliographies by Dietz (1944, 1956) or the general reference by Mantell (1951).

Extraction may also be accomplished by bubbling the air stream through a solvent. Water is often used as the solvent, although there is no reason not to use low-vapor-pressure hydrocarbons or other liquids for trapping specific substances. For quantitative work, it is frequently necessary to stabilize the species of interest so that it will not reenter the gas stream. Thus, sulfur dioxide may be stabilized as the dichlorosulfitomercurate ion, or it may be oxidized to sulfate ion with hydrogen peroxide. If qualitative analyses are to be performed, it may not be desirable to chemically change the species of interest during the collection operation.

Figure 3.6 shows some of the extraction devices in use. Greenburg–Smith and midget impingers are often used as extractors, although they are generally less efficient than fritted glass devices which generate smaller bubbles. Roberts and McKee (1959) investigated the efficiencies of several devices for the extraction of ammonia from air. Some of these efficiencies are shown in Fig. 3.7.

It should be noted that the midget impinger cannot be used with flow rates greater than 7 liter/min due to entrainment of the liquid in the air stream. The efficiency will depend on the gas being collected, the absorbing liquid, and the maximum pore size of the frit, in addition to the flow rate. Efficiencies greater than 99% have been reported for many gas–liquid combinations. The efficiency of a given device relative to the amount it is capable of collecting may be easily checked by connecting two extractors in series and measuring the amount of gas extracted by the second device. While a number of bubbler designs have been used and are available commercially, one of the more commonly used configurations is shown in Fig. 3.8. The design minimizes blowout of the absorbing liquid at higher flow rates. Axelrod *et al.* (1971) have recommended the use of a more durable version of this extractor in which the rather fragile glass upper unit is partly made of Teflon.

Spiral absorbing tubes, such as that shown in Fig. 3.9, are used in a number of automatic instruments. In these devices, the liquid metered in at the bottom of the tube is carried to the top of the spiral. The air stream

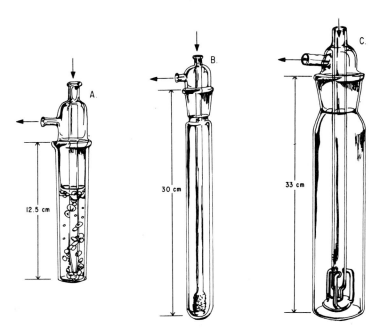

FIG. 3.6.   Gas absorption bottles. A. Midget impinger; B. fritted glass extractor; C. Greenburg–Smith impinger.

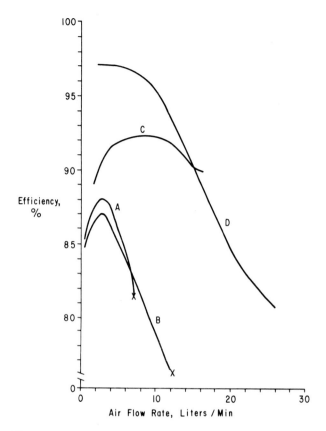

Fig. 3.7. Extraction efficiencies for 1 ppm ammonia at about 25°C with selected absorption bottles. A, Midget impinger; B, midget bubbler; C, Greenburg–Smith impinger; D, 250-ml gas washing bottle. (Roberts and McKee, 1959, reprinted by permission of the Air Pollution Control Association.)

enters at the bottom and the bubbles of air rising up through the spiral ensure reasonably efficient mixing of the liquid and gas phases.

3.2.3 COLLECTION OF PARTICLES

A variety of situations arise where sampling with filters is useful. The types of filters appropriate for a given application are determined by the sensitivity of the analytical method for the substance of interest. For example, sampling for atmospheric sulfate to be analyzed by conventional wet chemical techniques requires a large sample on a filter which has a low sulfate content and adequate retention characteristics for the particles, usually in the 0.1–1.0-μm diameter range. Low-sulfate fiber filters are usually chosen for their low flow resistance (ΔP), low price, and adaptability to the high-volume air sampler. The variety of filters available and some of their characteristics are shown in Table 3.1.

The theories of filter efficiency have been considered by Pich (1966) for fiber filters and membrane filters. While these theories are complex and approximate, they do provide insight into the mechanisms by which filters operate. In both types of filter, there are several processes by which particles are trapped. Some of these processes will be discussed individually below.

As will be shown in Chapter 9, the Stokes number, Stk, governs the

FIG. 3.8.   Gas absorption bottle designed to minimize entrainment of solvent.

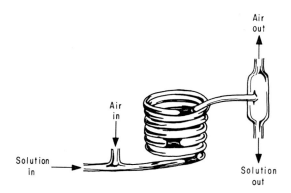

Air
out

Air
in

Solution
in

Solution
out

FIG. 3.9.   Spiral gas absorber of the type often used in automatic instrumentation.

efficiency of collection by impaction. Large, dense particles exhibit inertial deviations from the flow streamlines when the flow is forced to go around an object. If the particles strike the filter material as a result, they are removed from the air stream. Impaction is generally effective only for large, dense particles and becomes less effective as particle size or density decreases. High-velocity flow (therefore high particle velocity) favors impaction, while low velocity decreases its efficiency.

Very small particles exhibit sufficient Brownian motion to diffuse over distances of the order of a few tens of microns in times typical of the residence time of the air sample within the filter medium. This mechanism is important for both fiber filters and membrane filters and is favored for low flow rates and small particle sizes.

Particles which are larger than the space between fibers or larger than the pores of a membrane filter will obviously be trapped. While it may not be obvious, both the impaction and diffusion mechanisms effectively overcome the need for collection by interception. Nonetheless, the "sieve" effect is useful for the collection of very large particles.

If a particle is charged opposite to the charge of the filter medium, attractive forces will exist which enhance the above mechanisms and increase the efficiency. Little is known about the generalities of this mechanism.

Chemical filter papers, such as Whatman papers, have been used and, although their low ash content is an advantage for certain chemical analyses, they are relatively inefficient over a wide range of particle sizes and are difficult to use if one wishes to examine the aerosol by optical means. The properties of a variety of paper filters have been investigated by Smith and Surprenant (1954).

TABLE 3.1

SELECTED DATA ON FILTERS USED IN ATMOSPHERIC AEROSOL SAMPLING

Filter	Mean pore diameter,[a] μm	Air flow rate[b,c]	Thickness,[b] μm	Efficiency, %, at 5 cm/sec face velocity Particle diameter	
				0.03 μm	0.3 μm
Gelman membrane:					
GA-1	5.0	20	140	—	—
GA-3	1.2	18	140	—	—
AN800	0.8	13	140	99.3[a]	97.0[a]
GA-6	0.45	6.6	140	>99[d]	>97[d]
GA-7	0.3	4.2	140	>99[d]	>97[d]
GA-8	0.2	3.9	140	>99[d]	>97[d]
Millipore membrane:					
SC	8.0	55	125–150	—	—
SM	5.0	35	125–150	—	—
SS	3.0	20	125–150	99.9[a]	91.2[a]
RA	1.2	15	125–150	—	—
AA	0.8	11	125–150	99.9[a]	97.3[a]
DA	0.65	10	125–150	—	—
HA	0.45	4	125–150	99.9[a]	98.8[a]
PH	0.30	3.7	125–150	>99[d]	>99[d]
GS	0.22	2.5	125–150	>99[d]	>99[d]
VC	0.10	0.49	125–150	>99[d]	>99[d]
VM	0.05	0.31	125–150	>99[d]	>99[d]
VS	0.025	0.22	125–150	>99[d]	>99[d]
OH	1.5	—	125–150	87.2[a]	51.1[a]
OS	10.0	—	125–150	63.4[a]	31.2[a]
Nuclepore	8.0	~100[c]	10	5.7[a]	10.1[a]
Nuclepore	5.0	~100[c]	10	18.4[a]	14.4[a]
Nuclepore	3.0	>70	10	—	—
Nuclepore	2.0	~50[c]	10	43.3[a]	28.3[a]
Nuclepore	1.0	30	10	86.8[a]	52.2[a]
Nuclepore	0.8	22	10	94.6[a]	61.9[a]
Nuclepore	0.6	~10	10	—	—
Nuclepore	0.5	~10	10	98.7[a]	99.3[a]
Nuclepore	0.4	~10	10	>99[d]	>99[d]
Nuclepore	0.2	~4	10	>99[d]	>99[d]
Nuclepore	0.1	0.8	10	>99[d]	>99[d]
Fiber:					
A Gelman	NA[e]	—	~450	>98[b]	>99.7[b]
E Gelman	NA[e]	—	~450	>98[b]	>99.7[b]
PF-41 Whatman	NA[e]	—	~300[d]	58.9[a]	22.4[a]

[a] Spurney et al. (1969a,b).
[b] Data supplied by manufacturers.
[c] Flow rate in liter/min-cm² ($\Delta P = 0.93$ atm at 1 atm).
[d] Estimated.
[e] NA: Not applicable.

Glass fiber filters are available from several of the manufacturers associated with paper filters. These filters are available with very high efficiencies and may be used at temperatures up to 500°C. Although they are useful for many chemical and gravimetric analyses, they are not preferred for those cases in which the aerosol is to be examined optically.

The result of the mechanisms of impaction, diffusion, and interception is that the collection efficiency of a given filter as a function of particle size will be similar to that expressed in Fig. 3.10. The position and depth of the minimum will depend on the filter material and the stream velocity through the filter.

Membrane filters are made from a variety of materials, including cellulose esters, Nylon, Teflon, and polyvinylchloride. They are available with many effective pore sizes and the efficiencies are generally greater than those of paper filters. Most of them do not absorb water, although they may be dissolved in certain organic solvents (Gelman, 1965).

Nuclepore filters are made by subjecting a thin film of polycarbonate to a beam of atomic particles. The damaged material is then etched away by an appropriate solvent and the result is a filter having a uniform and controllable pore size. The theory and some applications of these filters have been investigated by Spurney et al. (1969a,b). Although Nuclepore filters are somewhat less retentive than membrane filters, their surface

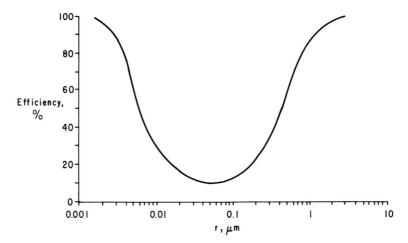

FIG. 3.10.   Dependence of filter efficiency on particle radius for a filter with a pore size of 2.5 μm. Spurney et al., (1969a), reprinted from Environ. Sci. Technol. 3, 453 (1969). Copyright 1969 by the American Chemical Society. Reprinted by permission of the copyright owner.

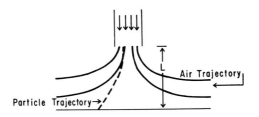

Fig. 3.11. Single-stage impactor.

uniformity makes them superior for use in optical or electron microscopy. The uniformity of pore size even allows them to be used for the sizing of aerosols in some cases. Nuclepore filters have a low mass per unit area and thus have an advantage for gravimetric determinations of aerosol concentration. They are reasonably nonhygroscopic and are soluble in chloroform.

It is difficult to generalize about the relative advantages of various filter media. Fuchs (1964) has defined a figure of merit, $\beta = \Delta P^{-1} \log(1 - \epsilon)$, where ϵ is the filter efficiency and ΔP is the pressure drop across the filter. The figure of merit β depends on particle size and density and the stream velocity through the filter. In view of the difficulties involved in defining filter performance, one should recognize the limitations of the filter used in each application.

Impactors, or impingers, collect aerosols by the mechanism of impaction. The operation of a single-stage impactor is illustrated in Fig. 3.11. The air stream at a velocity V_0 is directed at a flat plate a distince L from the orifice. The inertial forces tend to make the particle move in a straight line, while the drag forces tend to accelerate the particle in the direction of fluid motion. For a given size particle, the collection efficiency depends on the ratio V_0/L, increasing as V_0/L increases. Cascade impactors use a number of jets, with varying values of V_0 and L, connected in series. In the first jet, V_0/L is small, so that only the largest particles will be collected. In succeeding stages, V_0/L is increased so that successively smaller particles are collected.

There are a number of models of cascade impactors available, two of which are shown in Fig. 3.12. The collecting surface may be coated with a thin film of inert oil or grease to improve adhesion of the particles.

Portable electrostatic precipitators are available which operate on the same principle as the large industrial units. Particles passing between two electrodes with a sufficient potential to maintain a corona discharge become ionized and are collected on the surface of one of the electrodes.

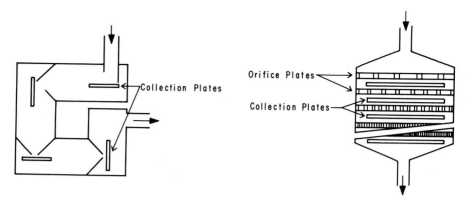

Orifice Plates

Collection Plates

Collection Plates

FIG. 3.12.   Two types of series impactors.

Electrostatic precipitators are quite efficient for collecting large particles. Although the efficiency may be improved by increasing the length of the electric field, the efficiencies for most working models are significantly less than unity for submicron particles. Electrostatic precipitators have been used for eliminating interferences from aerosols in gas analyses; however, it should be noted that the ozone and nitric oxide produced in negative discharge precipitators may cause chemical interferences.

Aerosols may also be affected by the forces of diffusiophoresis, photophoresis, and thermophoresis. These effects are discussed by Fuchs (1964) and Davies (1966). The only method which might have some potential in aerosol sampling is thermophoresis. Although the flow capacities of thermal precipitators are generally too low to permit the collection of aerosol samples by this method, they may be used to eliminate aerosols from gas samples. As in the case of electrostatic precipitators, one must take pains to ensure that the precipitator does not affect the composition of the gas sample.

3.3 Sampling Trains

It has already been noted that if the constituent is removed from the air, one must know the amount of air sampled in order to determine the concentration. For some methods, it will only be necessary to have the flow rate held at a constant value; for others, the absolute volume of the air samples will have to be measured. Detailed recommendations for the sampling of specific substances have been made by governmental agencies (Environmental Protection Agency, 1971; U.S. Dept. of Health, Education, and Welfare, 1965). We shall consider some general aspects of sampling in the following discussion.

3.3.1 MEASUREMENT OF VOLUME

Rates of flow may be measured by pitot tubes, orifice flowmeters, and rotameters. Diagrams of an orifice flowmeter and a rotameter are shown in Fig. 3.13. In the orifice flowmeter and the pitot tube, the flow rate is a function of a pressure change. In a rotameter the flow rate is a function of the position of the weighted bob suspended in the air stream. The flow measurements obtained with these devices must be combined with a measurement of time in order to calculate the volume of the air sample. If the flow rate is not constant, the total volume must be obtained from $V = \int_0^t (dV/dt)dt$. Some flowmeters are constructed in such a way that the flow rate is held constant.* Although calibration charts are usually supplied by the manufacturers of the flowmeters, this calibration should be checked using one of the methods to be described for the measurement of volume.

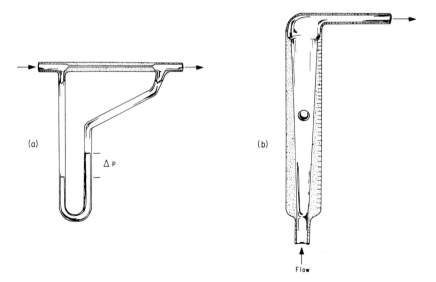

FIG. 3.13. Flowmeters. In the orifice flowmeter (a), the flow rate is a function of the pressure drop ΔP. The position of the bob is a measure of the flowrate in the rotameter (b).

* An example of this is the use of a hypodermic needle as a critical orifice as described by Lodge *et al.* (1966).

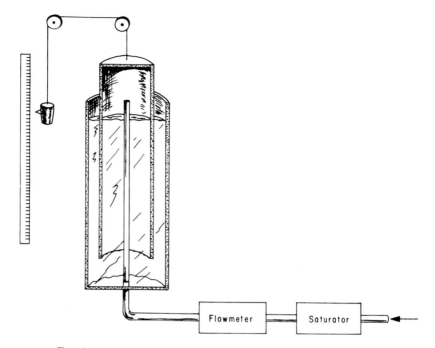

Fig. 3.14.   Apparatus for the absolute calibration of a flowmeter.

There are a number of instruments available which measure the volume of gas sampled. Dry test meters, wet test meters, and cycloid meters all operate by similar principles; they are available with direct readout in cubic feet or in metric units. It might be noted that gas meters which measure the volume in cubic feet may often be obtained at a lower cost and more conveniently than those calibrated in the metric system. One may calibrate a volume meter or flowmeter with a gas meter which has been calibrated recently. In some areas, public utility companies may calibrate gas meters for a nominal charge. The two methods described here have also been used for calibration.

The displacement of a liquid, such as water, provides one of the most direct methods of calibration. In a setup such as the one shown in Fig. 3.14, the pressure at the outlet of the gas meter remains constant. The air stream must be saturated with water at the temperature of the system. The volume of air is then determined from the change in position of the inverted container.

The soap bubble device shown in Fig. 3.15 provides a simple method

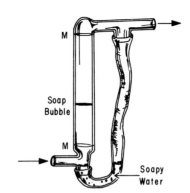

FIG. 3.15. Soap-bubble flowmeter. The length of time it takes for the soap bubble to pass between the two marks (M) is measured.

for measuring low flow rates. The rate is determined by measuring the length of time required for the soap bubble to travel the distance between the calibration marks.

3.3.2   SELECTED SAMPLING TRAINS

Before considering the details of sampling trains, we should consider what variables must be measured in order to calculate the concentration of a substance in the atmosphere. The following discussion assumes that the volume occupied by a substance A is very small with respect to the volume of air in which it is contained. If A is a gas, the concentration may be expressed as parts per million (by volume) defined by

$$c_A \text{ (ppm)} = 10^6 V_A / V_{air} \tag{3.1}$$

where the two volumes are measured at the same temperature and pressure. Normally, V_A will be obtained from a measurement of the number of moles of A, n_A. The value of V_{air} will be measured directly. Assuming that the ideal gas equation may be used,

$$c_A \text{ (ppm)} = 10^6 n_A RT / PV_{air} \tag{3.2}$$

where P and T are the pressure and temperature at the point where the volume of the air sample is measured. In some systems, one is safe in approximating the pressure in the flowmeter as the pressure of the atmosphere. Systems with filters, impactors, long inlet lines, or constrictions may operate with a significant pressure drop, in which case the pressure at the flowmeter must be known or controlled.

The concentration when A is a gas or particles may be expressed in

$\mu g(A)/m^3 (air)$ defined by

$$c_A' \ (\mu g/m^3) \ = \ 10^6 m_A/V_s \ = \ 10^6 n_A M_A/V_s \qquad (3.3)$$

M_A is the molecular weight of A and m_A is the weight of A in the sample of air. V_s is the volume of air containing the sample referred to some arbitrarily chosen temperature and pressure. While the pressure chosen is nearly always 1 atm, there does not seem to be any agreement on the choice of temperature. Many beginning chemistry students are familiar with the use of 0°C ($=273.15$ K) as a standard temperature; many thermodynamic functions are referenced to 25°C; the temperature of the meteorologist's standard atmosphere at sea level is 15°C. The use of 25°C has been recommended by the Environmental Protection Agency (1971). For some purposes, as in stack sampling, it is important to reference the concentrations to the environment from which the sample is collected. Needless to say, it is important to specify the conditions under which concentrations in $\mu g/m^3$ are measured. (Note that this problem does not exist when concentrations are expressed as ppm.) If T and P again refer to the temperature and pressure at the point of volume measurement and P_s and T_s are the standard pressure and temperature (or specify the conditions for which the concentration is desired) and V_{air} is the volume measured by the flowmeter, then

$$c_A' \ (\mu g/m^3) \ = \ 10^6 m_A P_s T/V_{air} P T_s \qquad (3.4)$$

Thus we can see that, in addition to measuring the amount of substance and volume of air, it is necessary to measure the temperature and pressure of the air at the flowmeter in order to calculate the concentration in ppm or $\mu g/m^3$. If, in addition, we wish to know the concentration in $\mu g/m^3$ in a stack, then we must also measure the temperature and pressure of air in the stack.

A sampling train may contain components which do the following: collect the substance of interest; remove interfering substances; move the air stream through the train; measure the temperature and pressure of the air stream; and regulate and measure the volume of the air. There are a number of ways to arrange these components; the final choice may be determined by the nature and reactivity of the constituent. Let us consider the sampling train shown in Fig. 3.16. If the pressure drop across the interference absorber and the extraction train is known to be small, it may be possible to approximate the pressure at the flowmeter by that of the atmosphere. This approximation may not be made when the extraction is carried out with systems employing large pressure drops, such as impactors and filters, or if the flow-limiting orifice is placed ahead of the flowmeter. It

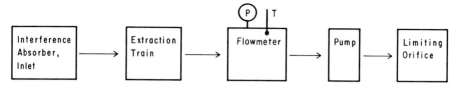

FIG. 3.16. Sampling train, example I.

should also be noted that Eqs. (3.2) and (3.4) assume that the pressure at the flowmeter is constant. This assumption may not be justified in applications where the impedence of the filter increases during the collection period. Assuming that the temperature of the flowmeter remains constant, the term PV_{air} must be replaced by $\int_0^t d(PV_{air})$ or $\int_0^t [d(PV_{air})/dt]\, dt$, depending on whether the actual volume or the flow rate is measured. These problems may be circumvented by changing the sample train as shown in Fig. 3.17. In this system, the pressure will be constant as long as the pressure drop across the flow-limiting orifice is constant. It should be clear that one must take precautions to ensure that the entire system is gas-tight up to the point where the volume is measured. The flowmeter and pump could also be placed near the inlet; however, this is not usually done, because of the possibility of interference with substances of interest.

In many instrument systems, the flow rates, pressure, and temperature are simply held constant and not explicitly measured. For such systems, it is only necessary to calibrate the instrument with a known gas mixture.

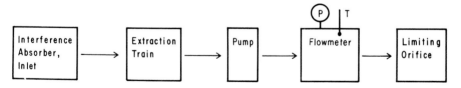

FIG. 3.17. Sampling train, example II.

3.3.3 CALIBRATION

All methods of analysis which measure something other than the actual mass of a substance, or number of particles, must be calibrated with a gas mixture of known composition. We shall discuss three methods of generating gas mixtures of known composition.

In the static dilution method, a known amount of the substance of interest is added to a known volume of air. The amount of substance may be determined from pressure–volume–temperature measurements, or mass measurements. Frequently, the dilutions must be carried out in two or more stages in order to obtain concentrations at the ppm level. The gas mixture may be stored in a cylinder under pressure. The method of storage may be determined by the type of analytical instrument used and the volume of gas required for calibration.

The dynamic method was proposed by Lovelock (1960) for the calibration of gas chromatographs. The dilution apparatus is shown in Fig. 3.18. If the initial concentration of A in the flask is N_0, the concentration of A reaching the detector as a function of time is given by

$$N_A = N_0 e^{-Ut/V} \qquad (3.5)$$

where V is the volume in which the mixing occurs and U is the flow rate of pure air into the mixing flask. This method is useful when the instrumental response time is short with respect to V/U and when it is necessary to calibrate the instrument over a wide range of concentrations. The method still requires some way of measuring N_0 unless the instrumental response at some concentration is known beforehand.

Permeation tubes were first developed by O'Keeffe and Ortman (1966, 1967) for the calibration of air pollution instruments. This method is probably the best method of calibration for those substances which can be contained in permeation tubes. A cylinder of a permeable material, such as FEP Teflon, is partly filled with the substance of interest in the liquid

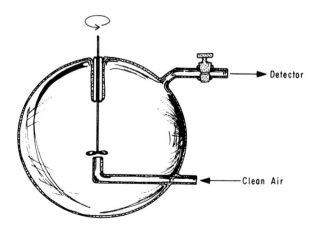

FIG. 3.18. Dynamic dilution apparatus. The concentration of substance initially present in the mixing flask decreases as pure air is passed through the system.

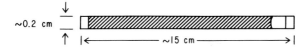

FIG. 3.19. Permeation tube.

state and sealed at both ends as shown in Fig. 3.19. The rate of loss of gas from the tube depends on the difference between the pressure of the substance inside the tube and the pressure outside the tube. The pressure outside the tube is kept at essentially zero by sweeping an air stream past the tube. In ideal cases, if there is always some liquid in the tube, the pressure in the tube will depend only on the temperature and not on the past history of the tube. Saltzman *et al.* (1971) have discussed some problems which arise in the use of permeation tubes for nitrogen dioxide and some hydrocarbons. The rate of loss of gas from the tube may be determined by weighing the tube at two times or by measuring the volume of gas emitted by the tube as a function of time (Saltzman *et al.*, 1969). An air stream containing a known concentration of a substance may then be prepared by passing air at a measured rate over a permeation tube maintained at a known temperature in a system similar to that shown in Fig. 3.20. The concentration of the substance may be varied by changing the flow rate of the clean air, the temperature, or the dimensions or character of the permeation tube.

Systems which analyze gas streams by measuring properties of the absorbing solutions may be calibrated by adding known amounts of the solute species of the gas. For example, NO_2 may be analyzed by absorbing it in an aqueous solution and forming the NO_2^- ion, which then undergoes subsequent reactions. This system may be calibrated by measuring the instrument response to known amounts of NO_2^-. In order to apply the results to the analysis of NO_2, assumptions must be made regarding the efficiency of collection and reduction of NO_2 gas.

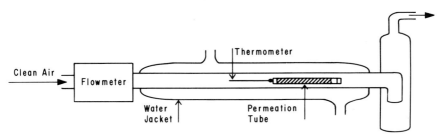

FIG. 3.20. Preparation of gas mixtures of known concentration with a permeation tube.

PROBLEMS

1. In an analysis of total particulate, a sample was collected on a Nucle-pore filter which weighed 0.0058 g before sampling and 0.0070 g after sampling. The uncertainty in each of the weighings was estimated to be 0.0002 g. The volume of air measured with a dry test meter was 400 ft³ at 30°C. The atmospheric pressure was 770 Torr and the pressure drop between the atmosphere and the dry test meter was 150 Torr. Ambient temperature was 15°C. Calculate the particulate concentration and associated uncertainty in $\mu g/m^3$ for ambient conditions and for "standard" conditions of 760 Torr, 25°C.

2. A sample train similar to that shown in Fig. 3.16 is used to measure the mass concentration of particulate matter by filtration. At an atmospheric pressure of 760 Torr, the pressure drop across the filter is 100 Torr. What will the percentage error in the mass concentration be if this pressure drop is neglected? Will the effect of this error increase or decrease the apparent mass concentration?

3. With the same sampling train as in Problem 2, what will the relative error be if the temperature of the flowmeter is 35°C and the temperature effect is overlooked in reporting the mass concentration at a reference temperature of 25°C?

4. The concentration of SO_2 in a stack gas was measured to be 500 ppm or 920 mg/m³ at 1 atm and 150°C. Calculate the concentration in ppm and mg/m³ at 1 atm and 25°C.

REFERENCES

Adams, D. F., Koppe, R. K., and Jungroth, D. M. (1960). *Tappi* **43**, 602.

Altshuller, A. P., Wartburg, A. F., Cohen, I. R., and Sleva, S. F. (1962). *Air Water Pollut.* **6**, 75.

Altshuller, A. P., Kopczynski, S. L., Lonneman, W. A., and Sutterfield, F. D. (1970). *Environ. Sci. Technol.* **4**, 503.

Axelrod, H. D., Wartburg, A. F., Teck, R. J., and Lodge, J. P. (1971). *Anal. Chem* .**43**, 1916.

Barrett, E. W., and Ben-Dov, O. (1967). *J. Appl. Meteorol.* **6**, 500.

Butcher, S. S., and Ruff, R. E. (1971). *Anal. Chem.* **43**, 1890.

Charlson, R. J., Ahlquist, N. C., Selvidge, H., and MacCready, P. B. (1969). *J. Air Pollut. Contr. Ass.* **19**, 937.

Craw, A. R. (1970). *Nat. Bur. Stand. Rep.* No. 10284.

Davies, C. N., ed. (1966). "Aerosol Science." Academic Press, New York,

Dietz, V. R. (1944). "Bibliography of Solid Adsorbants." Nat. Bur. Stand., Washington, D.C.

Dietz, V. R. (1956). Bibliography of solid adsorbants. *Nat. Bur. Stand. (U.S.) Circ.* No. **566**.

Environmental Protection Agency (1971). *Fed. Regist.* **36,** 8186.

Fuchs, N. (1964). "The Mechanics of Aerosols." Pergamon, Oxford.

Gelman, C. (1965). *Anal. Chem.* **37** (6), 29A.

Hoshino, H., Wasada, N., and Tsuchiya, T. (1964). *Bull. Chem. Soc. Jap.* **37,** 1310.

Jolly, W. L. (1960). "Synthetic Inorganic Chemistry." Prentice-Hall, Englewood Cliffs, New Jersey.

Leavitt, J. M., Pooler, F., and Wanta, R. C. (1957). *J. Air Pollut. Contr. Ass.* **7,** 211.

Lodge, J. P., Pate, J. B., Ammons, B. E., and Swanson, G. A. (1966). *J. Air Pollut. Contr. Ass.* **16,** 197.

Lovelock, J. (1960). *In* "Gas Chromatography 1960" (R. P. W. Scott, ed.). Butterworth, London.

Mantell, C. L. (1951). "Adsorption." McGraw-Hill, New York.

Moffat, A. J., and Millan, M. M. (1971). *Atmos. Environ.* **5,** 677.

O'Keeffe, A. E., and Ortman, G. C. (1966). *Anal. Chem.* **38,** 760.

O'Keeffe, A. E., and Ortman, G. C. (1967). *Anal. Chem.* **39,** 1047.

Pich, J. (1966). *In* "Aerosol Science" (C. N. Davies, ed.). Academic Press, New York.

Quiram, E. R., and Biller, W. F. (1958). *Anal. Chem.* **30,** 1166.

Quiram, E. R., Metro, S. J., and Lewis, J. B. (1954). *Anal. Chem.* **26,** 352.

Rasmussen, R. (1970). *Meeting Amer. Chem. Soc., 159th* (Div. of Air, Water, and Waste Chem.), *Houston, 1970.*

Roberts, L. R., and McKee, H. C. (1959). *J. Air Pollut. Contr. Ass.* **9,** 51.

Saltzman, B. E., Feldmann, C. R., and O'Keeffe, A. E. (1969). *Environ. Sci. Technol.* **3,** 1275.

Saltzman, B. E., Burg, W. R., and Ramaswamy, G. (1971). *Environ. Sci. Technol.* **5,** 1121.

Shepherd, M., Rock, S. M., Howard, R., and Stormes, J. (1951). *Anal. Chem.* **23,** 1431.

Smith, W. J., and Surprenant, N. F. (1954). *Amer. Soc. Test. Mater. Proc.* **53,** 1122.

Spurney, K. R., Lodge, J. P., Frank, E. R., and Sheesley, D. C. (1969a). *Environ. Sci. Technol.* **3,** 453.

Spurney, K. R., Lodge, J. P., Frank, E. R., and Sheesley, D. C. (1969b). *Environ. Sci. Technol.* **3,** 464.

Stalker, W. W., Dickerson, R. C., and Kramer, G. D. (1962). *J. Air Pollut. Contr. Ass.* **12,** 361.

U.S. Bur. of Mines (1967). Ringelmann smoke chart. *U.S. Bur. Mines Inform. Cir.* **IC 8333.**

U.S. Dept. of Health, Education, and Welfare (1965). Selected methods for the measurement of air pollutants. Publ. 999-AP-11. Pub. Health Serv., Cincinnati, Ohio.

Yamada, V. M., and Charlson, R. J. (1969). *Environ. Sci. Technol.* **3,** 483.

CHAPTER 4

TREATMENT OF DATA

The number which is supposed to represent the amount of contaminant in the air is affected by two fairly independent variables. The most obvious is the actual quality of air going into the detection and measuring device; the second is what the instrument does to the air sample in order to put out a number which we interpret as a measure of air quality. In most applications, it is assumed that variations in the number put out by the instrument are due solely to variations in air quality; however, somewhere along the line, one must make an evaluation of just what variations are introduced by the instrument itself. Once we understand the limitations of each analytical technique, we can get on with the problem of studying variations of atmospheric quantities. The other problem which will be considered in this chapter is the presentation of data.

4.1 Instrumental Error

Instrumental errors are usually classified into two types. *Systematic errors* are errors of more or less constant magnitude which result from the design of the particular procedure being used. Systematic errors may, in principle, be accounted for by careful examination of the analytical procedure and by conducting parallel experiments. In the Saltzman method for the analysis of NO_2 (Saltzman, 1954), this gas is absorbed from the atmosphere and then reduced to the nitrite ion, NO_2^-. The amount of NO_2^- is then determined by allowing it to react to form a dye, the concentration of which is obtained by measuring the absorbance and applying the Beer–Lambert–Bouguer law. It is generally more convenient to calibrate this method with a standard solution of $NaNO_2$ than to make up an air sample containing a known amount of NO_2. The casual use of the $NaNO_2$ method for standardization would, however, give rise to a rather large systematic error since it has been shown by Scaringelli et al. (1970) that under the conditions of analysis, one mole of NO_2 in the gas phase produces only 0.764 mole of NO_2^- in solution. (A systematic error may also be introduced by the use of the conversion factor 0.764, since subsequent work may show that this factor itself is in error or may depend on experimental conditions.) For this reason, it is important to specify the conversion factor used and the experimental conditions. Systematic errors may result from the following sources:

1. Calibration. The use of a calibration method which does not measure the same quantity as desired. Reagents of unknown quality may contribute to a systematic error.

2. Dirty inlet tubes. Partial removal of the species of interest before it gets to the instrument.

3. Interfering substances. These may increase or decrease the instrument response and give rise to a systematic error if they are present in a more or less constant amount.

Random errors comprise the "noise" of a strip chart record when the input is constant or the variations resulting from repeated independent measurements of the same quantity with the same instrument. The magnitude of the random error, or imprecision, depends on the method of measurement. The effects of random errors may be reduced by making a large number of measurements and averaging the results; however, it is usually useful to have some idea of the imprecision inherent in the method so that single measurements may be used.

A common measure of the magnitude of the random error in a single measurement is given by the standard error s, defined by the following equation for a case where we have n measurements of a variable x and obtain the values x_i:

$$s^2 = \sum_i (x_i - \bar{x})^2/(n-1) \qquad (4.1)$$

$\bar{x}$ is the mean or average value of x. A measure of the random error in the mean of n measurements, s_{mean}, is given by $s_{\text{mean}} = s/\sqrt{n}$.

The systematic and random errors must be combined to obtain a measure of the overall uncertainty of a measurement. Errors of measurement in a given system are not always fully described in the literature and an estimate must often be based to some extent on personal experience. The interested reader is directed to any general text on statistics. Recommendations for the description of uncertainties in report writing are contained in an article by Eisenhart (1968). Factors which must be considered in the evaluation of instrument performance are discussed in the articles by Rodes *et al.* (1969) and Palmer *et al.* (1969) for the case of SO_2-measuring instruments.

One factor which is related to error is the *sensitivity* of the method of analysis. By sensitivity is meant the change in a variable which will cause an instrument response of an arbitrarily chosen magnitude. The instrument response might be a change in transmittance from 100% to 99% in a photometric measurement. The sensitivity is often determined by extrapolation from measurements of a large instrument response. The sensitivity often places a lower limit on the uncertainty of a measurement.

The detection limit is the minimum value of concentration that causes an instrument response which may be distinguished from the zero concentration instrument response. If $\bar{x}$ is an average of instrument responses at some low concentration and $\bar{x}_0$ is the average response when the concentration is zero, then the minimum meaningful difference $\bar{x} - \bar{x}_0$ is equal to $\sqrt{2}Ks_0$, s_0 is the standard deviation of $\bar{x}_0$ and K is a constant which depends on the confidence limits chosen. If $\bar{x}_0$ is determined from a large number of measurements, $K = 1$ corresponds to 68% confidence limits, and $K = 1.96$

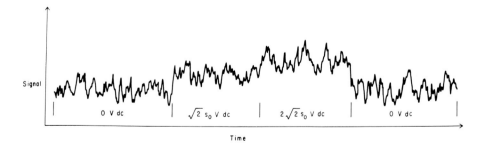

Signal

| 0 V dc $\sqrt{2}\,s_0$ V dc $2\sqrt{2}\,s_0$ V dc   0 V dc |

Time

FIG. 4.1.   Detection limits and random noise. The standard deviation, or root-mean-square noise, is s_0, indicated in the trace above 0 V dc. The 68% confidence level detection limit is indicated above $\sqrt{2}s_0$ V dc and the 95% confidence level detection limit is indicated above $2\sqrt{2}s_0$ V dc.

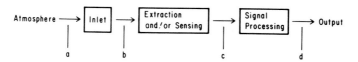

FIG. 4.2.   Idealized instrument system for the analysis of the time constants and residence times.

to 95% confidence limits. Choosing 95% confidence limits, the detection limit is then 2.8 s_0. The approximate relationships of $\bar{x}$, $\bar{x}_0$, and s_0 for different confidence limits are shown in Fig. 4.1. If the detection limit for a particular method of measuring SO_2 is, say, 0.02 ppm, a reported value of 0 ppm for SO_2 does not really mean 0.00 ppm, but rather that the concentration is less than 0.02 ppm.

Two other factors will be discussed which do not necessarily give rise to errors, but do place limitations on the interpretation of the data. The *time constant* and *residence time* are most easily discussed with reference to the following idealized example. Our instrument system is resolved into these components: inlet, extraction, and signal processing, as shown in Fig. 4.2. In some systems, the extraction step may be absent. We shall assume for the moment that our instrument system is free of interferences and that the time factors depend on the characteristics of the system and not on the properties of the substance being analyzed. Let us imagine that a sharp pulse of concentration of the substance occurs in the atmosphere at point a near the inlet at time zero as shown in Fig. 4.3. After a short length of time, the substance will be carried through the inlet. If we could imagine "seeing" the concentration at point b instantaneously, the pulse of concentration would appear as shown by the tracing b in Fig. 4.3. Actually, the event would not occur as the rectangular response shown, but this simple shape will be used to illustrate this example. The event at point b

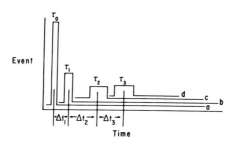

FIG. 4.3.   Signal pulse analysis.

will occur at a time Δt_1 after the event at point a. This is the transit time due to passage through the inlet tube. In addition, the pulse will be broadened due to diffusion and turbulence in the inlet. The increase in the event interval $\tau_1 - \tau_0$ is related to the time constant of the inlet. There will be another residence time Δt_2 and time constant $\tau_2 - \tau_1$ associated with the extraction step so that the event "seen" at point c would appear as shown by tracing c in Fig. 4.3. A final transit time and time constant are associated with the signal processing step, so that the information received by the instrument operator is shown by the tracing labeled d in Fig. 4.3. The time constant of the entire system is of the order of $\tau_3 - \tau_0$ and the residence time is Δt_{tot}. Although the magnitudes of some of these times in relation to properties of the instrument system are clear, it is best to discuss each of the contributions individually. The events at the four points are shown separated for clarity; in general, they may overlap.

In the inlet system, Δt_1 and $\tau_1 - \tau_0$ are functions of the size and shape of the inlet pipe and the flow velocity through the pipe. A quantitative treatment of these factors is available in the chemical engineering literature [see, for example, Himmelblau and Bischoff (1968)]. In remote measuring systems, Δt_1 approaches zero; in systems where the inlet pipe is several meters long, Δt_1 may be several tens of seconds. The extraction step might consist of absorption of a gaseous pollutant by a liquid, or filtration of an aerosol prior to weighing. If the extraction is done batchwise, as in filtration and noncontinuous extraction processes, $\tau_2 - \tau_1$ is the time interval over which the batch is extracted, and Δt_2 is the time required to extract the sample and prepare it for the final measurement. In continuous extraction processes, such as spiral absorbing tubes, $\tau_2 - \tau_1$ and Δt_2 are best determined empirically. The time Δt_2 is often of the order of several minutes and $\tau_2 - \tau_1$ is generally much shorter than Δt_2 for spiral absorbing tubes.

Signal processing is that part of the analysis during which the output is obtained while nothing happens to the sample chemically or physically. The signal processing may be entirely electronic in nature or it may be the actual weighing of a filter sample. If the processing is achieved electronically in real time, Δt_3 will be of the order of a few seconds or less. On the other hand, if this final step is carried out mechanically, Δt_3 may be of the order of days, but it may also be known with high accuracy. The reader will recognize that in some systems the inlet or the extraction step may be absent; in other systems, these components may be the most important in determining $\tau_3 - \tau_0$ and Δt_{tot}. It should also be recognized that there are limitations on what one may infer about the conditions in the atmosphere from the response of the instrument system, particularly when concentrations are changing rapidly.

4.2 Presentation of Data

Initially, data are usually presented as a function of time. All details of the analysis should also be available so that the person using the data will have a clear idea of their limitations. Uncertainties of the measurements, possible interferences, sensitivity, averaging time, and residence time should be made clear. The raw data may be presented in analog form, as a strip chart record, or in digital form if manual analyses have been performed. It is often interesting to see the time dependence of the concentrations in order to make traffic or weather correlation studies or to investigate the dependence of the concentration of one substance on that of another. In addition to the time dependence, the data are usually processed further in order to determine how frequently a certain concentration is reached, or what the probability is that a specified concentration will be exceeded in a given time interval.

In the following discussion, we shall be making use of the data given in analog form in Fig. 4.4 and in digital form in Table 4.1. Note that although the analog representation is continuous, we have taken 1-hr averages in forming the table. The strip chart records transmittance as a function of time. Transmittance may be converted to ppm or $\mu g/m^3$ manually or by means of an analog device on the analytical instrument. The data may be analyzed directly to obtain averages for the desired time intervals. We shall illustrate a few methods for representing the data in graphical form. The first step in the further treatment of the data in the table is to divide the concentration range into a number of class intervals and count the number of data points within each class interval. The number of intervals chosen will depend on the resolution required and on the computation

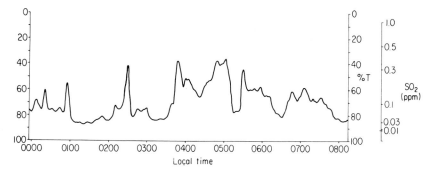

Fig. 4.4. Sulfur dioxide monitoring data in analog form. (Reproduced by permission of the Maine Environmental Improvement Commission.)

TABLE 4.1

HOURLY AVERAGES OF SO$_2$ CONCENTRATION (μg/m^3, 0°C)[a]

Hour	Station I				Station II			
	Day 1	Day 2	Day 3	Day 4	Day 1	Day 2	Day 3	Day 4
0000	200	430	340	230	134	236	272	86
0100	460	460	315	370	143	240	350	137
0200	400	485	890	315	202	350	329	109
0300	430	370	230	430	213	500	325	270
0400	1430	370	2120	430	179	468	282	282
0500	890	370	2120	370	168	375	257	386
0600	1040	340	2666	460	240	290	257	400
0700	515	286	1060	315	261	236	247	308
0800	260	230	545	315	229	206	447	193
0900	170	145	660	515	222	127	318	197
1000	145	116	1060	286	286	57	193	179
1100	145	116	600	260	200	36	119	96
1200	116	116	600	315	236	39	122	99
1300	87	116	485	315	315	24	86	236
1400	230	145	315	315	332	24	143	50
1500	170	145	260	286	458	11	109	75
1600	260	170	315	315	443	19	137	206
1700	260	230	340	315	397	92	140	340
1800	286	170	400	315	361	118	145	325
1900	230	145	340	340	382	147	204	429
2000	315	230	400	400	479	184	139	529
2100	370	260	400	545	454	229	122	483
2200	400	315	230	2140	279	183	106	468
2300	400	286	145	1090	247	175	212	615
Avg.	384	252	709	449	286	182	211	271

[a] Data courtesy of the Maine Environmental Improvement Commission.

facilities available. The size of all intervals should be the same. When this has been done, the data can be represented in tabular form or in graphical form as shown by the histogram in Fig. 4.5. This particular form is known as a frequency–concentration representation. The frequencies may also be normalized and given as percentages by dividing by the total number of data points.

The data may also be represented by plotting a graph of cumulative percentage against concentration. The cumulative percentage is the percentage of all cases having a concentration less than a certain value, which in this case has been taken as the upper limit of each class interval. The

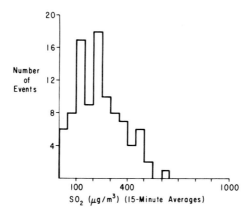

FIG. 4.5. Frequency–concentration histogram of the data for station I.

cumulative percentage–concentration plot is given for station I in Fig. 4.6.

Another representation which is often used is a concentration versus standard variable graph, or probability graph. In this graph, the concentration is plotted as the ordinate and the cumulative percentage, on a nonlinear scale, is plotted as the abscissa. The data points on such a graph will lie on a straight line if the data is normally distributed as given by the equation

$$f(x) = [1/\sigma(2\pi)^{1/2}] \exp[-(x - \bar{x})^2/2\sigma^2] \qquad (4.2)$$

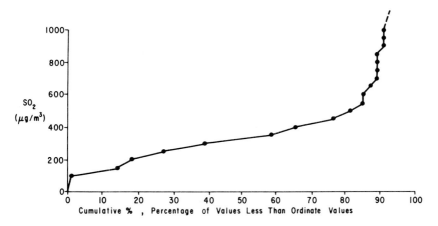

FIG. 4.6. Concentration–cumulative percentage plot for station I data.

where $f(x)$ is the fractional frequency per unit interval x, x is the concentration, $\bar{x}$ is the mean concentration, and σ is the standard deviation. In these graphs, the distance along the abscissa is measured in units of the standard variable $z\ [=\ (x - \bar{x})/\sigma]$, which is related to the cumulative percentage by the following expression for a normal distribution:

$$\text{cumulative } \%/100 = [1/(2\pi)^{1/2}] \int_{-\infty}^{z} \exp(-y^2/2)\, dy \qquad (4.3)$$

Such representations have the advantage that, if the distribution is normal, $\bar{x}$ is the value of x at the 50% point. The standard error is related to the slope and is the difference between the 50% cumulative concentration and the 84.13% cumulative concentration.

It has been observed for many variables that the logarithm of the variable is more nearly normally distributed than is the variable itself. That is, the distribution is given by the expression

$$F(x) = \left[\frac{1}{\ln \sigma_g (2\pi)^{1/2}}\right] \exp[-(\ln x - \ln x_g)^2/(2 \ln^2 \sigma_g)] \qquad (4.4)$$

$F(x)$ is the frequency per unit interval $\ln x$, x_g is the geometric mean, and σ_g is the standard geometric deviation. The frequency distribution of $\ln x$ may be obtained from the frequency distribution of x, without constructing class intervals of equal size in $\ln x$, by making use of the transformation $F(x) = xf(x)$. If one is constructing a $\ln x$ (or $\log x$) cumulative graph, it is not necessary to make this transformation. Data for stations I and II are represented on a log probability plot in Fig. 4.7. The

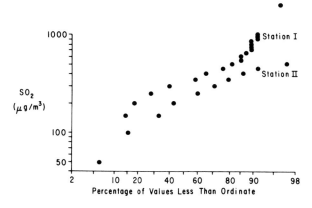

FIG. 4.7.   Log concentration–probability plot for the data in Table 4.1.

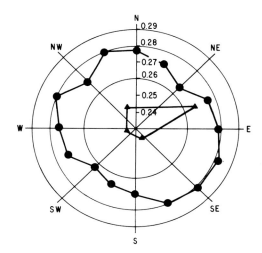

FIG. 4.8. N_2O and SO_2 concentrations as a function of wind direction for Mainz, Germany. Dots: N_2O concentration, scale given. Triangles: SO_2 concentration, linear scale, not given, with zero at the center. (Schütz *et al.*, 1970; reprinted by permission of the American Geophysical Union.)

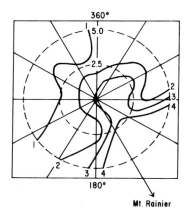

FIG. 4.9. Visibility of Mt. Rainier from Seattle as a function of wind velocity. The radial coordinate is wind speed (m/sec) and the angular coordinate is wind direction. The curved lines are isopleths of the scattering coefficient (in units of 10^{-4} m^{-1}) of the atmosphere in Seattle for conditions in which Mt. Rainier is barely visible from Seattle. (Ziegler *et al.*, 1971; reproduced by permission of the American Meteorological Society.)

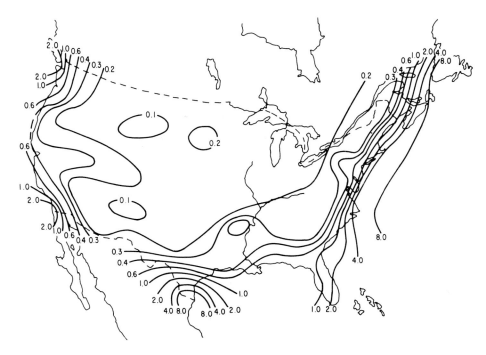

Fig. 4.10.   Average Cl⁻ concentration (μg/liter) in rain over the U.S. July 1955 to June 1956. (Junge and Werby, 1958; reprinted from the *Journal of Meteorology*, by permission of the American Meteorological Society.)

utility of the log normal distribution in describing air pollution data has been discussed by Larsen (1969).

Concentration and effects data may also be combined with meteorological data in order to obtain information on the effects of weather and possible sources of pollution. One representation which has been used is a plot of concentration as a function of wind direction on a radial diagram. In these plots, the concentrations are usually averages over a large number of wind speeds for a given wind direction. Figure 4.8 is such a plot for N_2O and SO_2 in Germany and shows that the N_2O concentration is quite insensitive to wind direction, relative to SO_2, and probably results from widespread natural sources.

Some indication of the dependence of an effect on wind velocity is also obtained if the effect is plotted as isopleths on a wind direction/wind speed rose as shown in Fig. 4.9. This figure illustrates that the visibility of Mt. Rainier from Seattle is poorer when the wind is blowing from the city to the mountain than when the wind is blowing from the mountain to the city (Ziegler *et al.*, 1971). The effect of dilution by increased wind speed for other wind directions may also be seen.

Concentration isopleths may also be drawn on maps of any scale. The concentrations may represent averages over a period of months or years or they may represent averages over a period when meteorological conditions are essentially constant. The concentration isopleths for chloride in rainwater obtained by Junge and Werby (1958) from a network of stations in the United States is an example of the former case (see Fig. 4.10). Here, the influence of the oceans is clearly illustrated. Concentration isopleths for short time averages have been used to evaluate plume models and the influence of meteorological conditions. Figure 4.11 shows the isopleths of light scattering obtained by mobile monitoring in a period of about 30 min in Tacoma, Washington (Lutrick, 1971). In this case, it may be seen that turbulence and topographic effects bring the plume from a

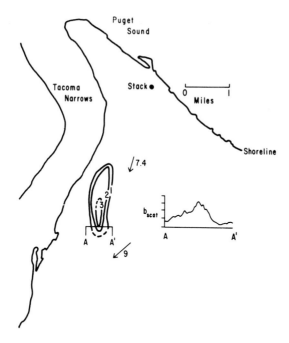

FIG. 4.11. Isopleths of light scattering coefficient at ground level measured with an integrating nephelometer in Tacoma, Washington, 14 August 1970, 1400–1415 PDT. Light scattering units are 10^{-4} m^{-1}, as discussed in Chapter 9. Data based on Lutrick (1971).

large stationary source to ground level about four miles from the source. The concentration may also be represented in a vertical plane if the altitude dependence of concentration is measured [see, for example, Johnson and Uthe (1971)].

We are not attempting to illustrate all possible ways of representing air monitoring data in this chapter. It should be recognized that some representations are more suitable for certain phenomena or characteristics and that no one method is best for all purposes. While certain standard methods will continue to be used, one should recognize the limitations of any particular representation and be willing to experiment with different forms.

PROBLEMS

1. Determine $\bar{r}$, σ, r_g, and σ_g for the aerosol size distribution given in the following tabular form $(1.12E + 05$ is 1.12×10^5, etc.)[a]:

Diameter interval (μm)	N (cm^{-3})	Diameter interval (μm)	N (cm^{-3})
0.0075–0.010	1.12E + 05	0.48–0.62	1.75E + 01
0.010–0.015	6.37E + 04	0.62–0.70	7.29
0.015–0.020	1.03E + 04	0.70–0.84	3.06
0.020–0.030	1.96E + 04	0.84–0.92	1.50
0.030–0.040	1.07E + 04	0.92–1.18	1.58
0.040–0.060	1.48E + 04	1.18–1.36	5.38E − 01
0.060–0.080	6.16E + 03	1.36–1.60	4.11E − 01
0.080–0.100	3.66E + 03	1.60–2.04	4.02E − 01
0.100–0.124	3.01E + 03	2.04–2.40	1.91E − 01
0.124–0.150	1.39E + 03	2.40–3.10	1.24E − 01
0.15–0.20	1.92E + 03	3.10–3.50	1.91E − 02
0.20–0.30	1.56E + 03	3.50–4.74	9.57E − 02
0.30–0.40	3.28E + 02	4.74–5.70	1.91E − 02
0.40–0.48	8.28E + 01	5.70–6.96	2.87E − 02

[a] Data courtesy of Prof. K. T. Whitby, University of Minnesota.

2. Two different instruments for the analysis of nitrogen oxides are attached to the end of an inlet tube 6 m long and 1.5 cm in diameter through which air is sampled at a rate of 5 liters/min. One instrument, which analyzes for NO_2 colorimetrically using a spiral absorbing tube, has an instrumental delay time of 8 min and an averaging time of 30 sec. The other instrument analyzes by the chemiluminescent method

and has a delay time of 10^{-2} sec and an averaging time of 10 sec. Discuss the factors limiting the data obtained by these instruments.

3. In an application of the Saltzman method for the determination of NO_2, the absorbance was measured of the dye formed from a given volume of a stream of 0.10 ppm NO_2. The absorbance values obtained were: 0.50, 0.54, 0.53, 0.48, 0.51, and 0.53. (a) Calculate the sensitivity, defined as the amount of NO_2 which will give a transmittance of 99%. (b) Calculate the 95% confidence level detection limit, assuming that the standard deviation of absorbance at 0 ppm is the same as that at 0.10 ppm.

REFERENCES

Eisenhart, C. (1968). *Science* **160,** 1201.

Himmelblau, D. M., and Bischoff, K. B. (1968). "Process Analysis and Simulation." Wiley, New York.

Johnson, W. B., and Uthe, E. E. (1971). *Atmos. Environ.* **5,** 703.

Junge, C. E., and Werby, R. T. (1958). *J. Meteorol.* **15,** 417.

Larsen, R. I. (1969). *J. Air Pollut. Contr. Ass.* **19,** 24.

Lutrick, D. (1971). M.S. Thesis, Univ. of Washington, Seattle.

Palmer, H. F., Rodes, C. E., and Nelson, C. J. (1969). *J. Air Pollut. Contr. Ass.* **19,** 778.

Rodes, C. E., Palmer, H. F., Elfers, L. A., and Norris, C. H. (1969). *J. Air Pollut. Contr. Ass.* **19,** 575.

Saltzman, B. E. (1954). *Anal. Chem.* **26,** 1949.

Scaringelli, F. P., Rosenberg, E., and Rehme, K. A. (1970). *Environ. Sci. Technol.* **4,** 924.

Schütz, K., Junge, C., Beck, R., and Albrecht, B. (1970). *J. Geophys. Res.* **75,** 2230.

Ziegler, C. S., Charlson, R. J., and Forler, S. H. (1971). *Weatherwise* **24,** 114.

CHAPTER 5

SPECIAL METHODS OF ANALYSIS

The purpose of this chapter is to provide enough information about selected methods of analysis so that the reader can gain some idea of the limitations and advantages of each method relative to his particular problem. This chapter does not pretend to provide all the details actually required to go out and use a method; the reader will have to go to some of the works indicated in the references.

5.1 Chromatography

Chromatography is basically a method for separating a mixture of compounds. The essential features of a chromatographic unit are shown in Fig. 5.1. The mobile phase is a gas or liquid which is inert with respect to the sample. It flows at a more or less constant rate in one direction through the stationary phase, which may be a solid with a large surface-to-volume ratio, or a high-boiling liquid on a solid support. The sample may be a gas, solid, or liquid; however, it must be soluble in the mobile phase. The term detector is used rather broadly here to include any unit capable of determining whether or not a substance is present in the mobile phase.

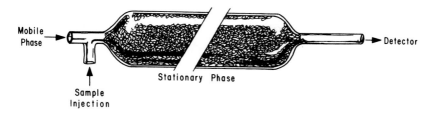

FIG. 5.1. Generalized chromatography apparatus.

The basis of the chromatographic separation is the partition ratio K, which describes the equilibrium between sample molecules in the mobile phase and sample molecules in the stationary phase. For substance A,

$$K_A = \frac{N_A \text{ per unit length in stationary phase}}{N_A \text{ per unit length in the mobile phase}} \tag{5.1}$$

If K_A is very small, A has little affinity for the stationary phase relative to the mobile phase and therefore moves through the stationary phase rapidly. As K_A becomes larger, the A molecules spend relatively more time attached to the stationary phase and thus move through the stationary phase more slowly. The length of time required for a substance to pass through the stationary phase is proportional to $1 + K_A$ and therefore substances with different partition ratios will require different lengths of time to pass through the system to the detector. In practice, the sample is injected as a "plug" and appears at the detector as a broadened plug as shown in Fig. 5.2.

Detectors range from high-resolution mass spectrometers coupled to computers, to very simple thermal conductivity cells, to the analyst's eye. The materials may be collected at the detector end and subjected to further separations or chemical tests.

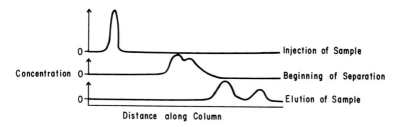

FIG. 5.2. Separation of components of a mixture as a function of time.

5.1.1 GAS CHROMATOGRAPHY

In gas chromatography, the mobile phase is an inert gas (most commonly, nitrogen or helium) and the stationary phase is either a high-boiling liquid supported on a porous solid or a solid contained in a tubular column. The choice of stationary phase will depend on the types of compounds being separated. Some stationary phases separate compounds according to volatility, while others are more sensitive to the presence of functional groups. A wide variety of columns may be made in a moderately equipped laboratory; a number of commercial columns are also available.

The detector is the heart of a gas chromatograph and although many types are available, we will discuss three types in common use to give some idea of the range and limitations of different detectors.

The thermal conductivity cell, or catharometer, detects the presence of a substance in the carrier gas by measuring the change in thermal conductivity. Thermal conductivity is mainly a function of molecular weight M and collision cross section A for a given molecule and is proportional to $A/M^{1/2}$ at a constant temperature. The larger the difference between the thermal conductivity of the sample and the mobile phase, the greater will be the sensitivity of the catharometer. For this reason, helium is often used as the mobile phase, although nitrogen may be used if the sample concentrations are large or the molecular weights are high. Catharometers may be applied to the detection of a broad range of gases. The limiting sensitivity depends on a large number of variables, but at best, is of the order of 1 ppm of substance in the sample.

The flame ionization detector, or FID, is one detector in a class of detectors which ionize the effluent gas by some process and then measure the electrical conductivity of the gas stream. In this detector, enough hydrogen gas is mixed with the effluent gas to maintain a steady hydrogen–oxygen or hydrogen–air flame. (Pure hydrogen may be used as the mobile phase.) Although relatively few ions are produced in a hydrogen flame, the presence of certain substances greatly increases the conductivity of the flame. The FID is very sensitive to hydrocarbons and to organic compounds containing a large number of CH groups. This detector is insensitive to many atmospheric constituents, including: the rare gases; nitrogen; the oxides of nitrogen, carbon, and sulfur; water; ammonia; and hydrogen sulfide. The response of the FID increases as the number of CH groups increases. The limiting sensitivity to propane of about 0.003 ppm in the total sample means that this detector can be used for the analysis of some constituents at atmospheric concentration levels.

The electron capture detector, ECD, detects the presence of certain substances by their ability to capture thermal electrons. A low-energy beta emitter, such as ^{63}Ni or ^3H, is used as the source and the capture of electrons results in a decrease in the conductivity of the gas because of the low mobility of the resulting ions. The response of the ECD is a very sensitive function of the molecular structure and it is difficult to generalize about specific sensitivities. The detector generally responds to molecules containing halogens or oxygen atoms and molecules with conjugated double bonds. (The specific responses to molecules in these classes varies over several orders of magnitude.) The ECD is insensitive to alkanes and the noble gases. In the usual mode of operation of the ECD, specific responses are very sensitive to temperature and other operating conditions and therefore a great deal of care must be exercised when using this detector for quantitative work. Lovelock *et al.* (1971) have described a method in which the ECD is used as an absolute gas phase coulometer, enabling accurate quantitative measurements of some substances to be made. The high sensitivity of this detector to sulfur hexafluoride (1×10^{-5} ppm in the sample) has permitted the use of SF_6 as an atmospheric tracer. The detection limit of the ECD for peroxyacetylnitrate (PAN) is about 5×10^{-3} ppm and thus low enough for use as a direct atmospheric monitor in certain cases (Darley *et al.*, 1963).

Compounds may be characterized by their retention times (the time required for the compound to reach the detector) or retention volume (the volume of carrier gas passing through the column during the retention time). Relative retention times are sensitive functions of the stationary–phase–sample combination. In the best of circumstances, all substances in a mixture have different retention times and this parameter and detector response serve to indicate the presence and quantity of a substance. There is always the chance that a given peak is due to an unexpected compound or perhaps a mixture of compounds. Since detector response and retention time do not provide definitive evidence as to the identity of the material being eluted, gas chromatographs are often coupled to mass spectrometers or other more specific analytical instruments. Details of some of the specialized methods used in conjunction with gas chromatographs are provided in the book by Ettre and McFadden (1969).

In favorable cases, the gas chromatograph can provide qualitative and quantitative information about a complex mixture. The use for quantitative analysis requires that the systems be calibrated with gas mixtures of known concentrations which span the concentrations actually being measured. The linearity of detector response with respect to concentration varies from one detector to another.

Where gas chromatography is useful for substances having appreciable vapor pressures at some temperature where decomposition does not occur, thin-layer and column chromatography may be used to separate involatile substances. In this method, the stationary phase is a porous solid and the mobile phase is a liquid.

5.1.2 THIN-LAYER CHROMATOGRAPHY

Thin-layer plates are formed by spreading a slurry of the desired solid on a glass plate and allowing the liquid to evaporate. Small amounts of the sample solution are spotted near an edge of the thin-layer plate and allowed to dry. The plates are then placed on edge in a shallow dish containing the mobile phase which passes through the porous solid by capillary action. When the solvent front reaches a point near the opposite edge of the plate, the separation is terminated by removing the plate from the dish containing the solvent. If the substances being separated are visible, they appear as spots at various positions on the plate as shown in Fig. 5.3.

The parameter analogous to retention time which is used to characterize substances is the R_F value, defined by

$$R_F = \frac{\text{distance moved by substance}}{\text{distance moved by solvent front}} \tag{5.2}$$

Special tools may be purchased or improvised for making thin-layer plates in the laboratory. Many commonly available solvents may be used as the mobile phase. Silica gel and alumina are widely used as the stationary phase. A variety of thin-layer plates are also available commercially. The performance of a solvent may depend on its purity, so that this variable must often be controlled. Mixtures of solvents are used in many applications.

The plates may be evaluated qualitatively by comparing the observed R_F values with the values for known substances determined from an independent series of tests. The R_F value depends on just about all variables

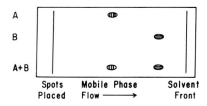

FIG. 5.3. Separation of a mixture by thin-layer chromatography.

imaginable, so that pains must be taken to ensure that the conditions under which the control substance was evaluated are identical to those for the unknown. The spots of most organic substances may be rendered visible by exposing the finished plate to iodine vapor or spraying it with a solution of potassium permanganate. The use of fluorescent indicators with examination by ultraviolet light is suitable for some materials. The spots should be located by a nondestructive method if the sample is to be subjected to further chemical tests.

Qualitative analyses may be confirmed with the use of the usual reagent tests and instrumental methods. A number of spot tests are described in the books by Fiegl (1966, 1958) for testing for inorganic ions and organic functional groups. If the substance is present in sufficient amounts, it is possible to scrape the spot area from the glass plate, extract the substance from the solid, and use any of the instrumental methods described elsewhere in this chapter.

The plates may be analyzed quantitatively by measuring the area or optical density of the spot. In either case, calibration curves must first be obtained using known amounts of the substance under identical conditions. The spot area and optical density measurements are simple to perform and instruments are available to make measurements rapidly; however, the results are generally subject to rather large uncertainties. More careful analysis requires the substance to be separated from the plate and measured quantitatively by other means. An example of the application of thin-layer chromatography to the analysis of several atmospheric polynuclear hydrocarbons is provided by Sawicki et al. (1964).

5.1.3 COLUMN CHROMATOGRAPHY

Column chromatography is similar to thin-layer chromatography with respect to applications and materials used. In this method, the solid adsorbent is contained in a vertical glass tube. The sample and the mobile liquid phase are added to the top of the column and eluted at the bottom. Samples of the liquid phase are collected at regular intervals and subjected to chemical and physical tests. Larger samples may be handled with this method than with thin-layer chromatography and it is somewhat easier to isolate the sample if further tests are necessary.

Samples analyzed by thin-layer or column chromatography may be obtained from the atmosphere as the high-boiling fractions of cold trap condensates or as filter extracts.

5.2 Spectrometry

Spectrometric methods of analysis make use of the discrete energy levels of molecules and the emission or absorption of radiation which usually accompanies changes by a molecule from one energy level to another. There are two types of spectroscopic experiments commonly used in analytical chemistry and they are illustrated in Fig. 5.4.

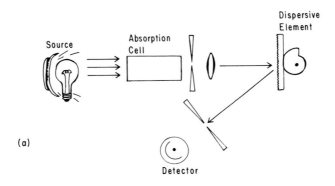

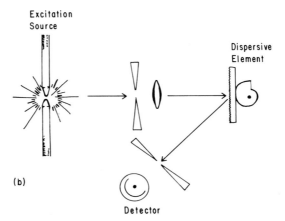

FIG. 5.4.   Basic spectroscopic experiments. (a) Absorption spectroscopy; (b) emission spectroscopy.

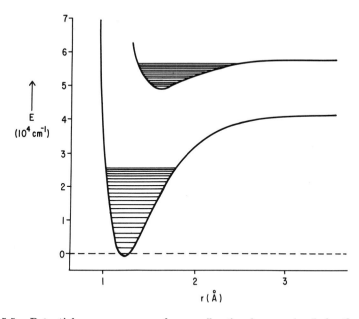

Fig. 5.5.   Potential energy curves and some vibrational energy levels for the ground state and an excited state of the O_2 molecule. [Herzberg (1950). Reproduced by permission of the Van Nostrand-Reinhold Company.]

In an absorption experiment, radiation from the source passes through the sample, where some of the energy is absorbed. The rest of the light is resolved into its constituent frequencies by a dispersing element, such as a grating or a prism, and then passes on to a detector. A comparison between the amount of light reaching the detector when the sample is present with that reaching the detector when the sample is absent is an indication of how much light is absorbed by the sample at that particular wavelength. In some instruments, the dispersion device precedes the sample, which is then illuminated by monochromatic radiation. The principle of operation, however, is the same.

In an emission experiment, an excitation source places the molecules in an excited (higher-energy) state. The molecules then emit light when they lose energy in dropping from the excited state to a lower-energy state. The emitted light is passed through a dispersion device and its intensity is measured by a detector. Exothermic chemical reactions (flames, chemical luminescence) and light (fluorescence) may serve as sources of excitation.

A partial energy level diagram for the oxygen molecule is shown in Fig. 5.5. The solid curves represent potential energy curves for different *elec-*

tronic states of this molecule. The fine horizontal lines represent the *vibrational* states for each of the electronic states. It is significant to note that the energy difference between adjacent vibrational states in a given electronic state is much less than the difference between the electronic states. We are not including the rotational states in this diagram. There will be a set of a few hundred rotational states associated with each vibrational state; the separation of the rotational states will be much smaller than the separation of the vibrational states. Molecules containing more atoms will in general have more electronic states and there will be more vibrational states associated with each electronic state.

Molecular transitions which may occur during the two spectroscopic experiments we have outlined are shown in Fig. 5.6. Absorption of a photon of light is accompanied by an *increase* in the energy of the molecule (from E_1 to E_2 or E_3). Emission of a photon occurs with a decrease in the energy of the molecule (from E_3 to E_2). The relationship between the change in energy and the frequency of light ν is given by the equation $E_{\text{upper}} - E_{\text{lower}} = h\nu$, where h is the Planck constant (6.6256×10^{-27} erg sec). Radiation is often characterized by its wavelength λ, which is related to the frequency by the relationship $\nu\lambda = c$, where c is speed of light (2.997925×10^{10} cm sec^{-1}). It can be seen that the energy change and the wavelength of light emitted or absorbed are inversely related.

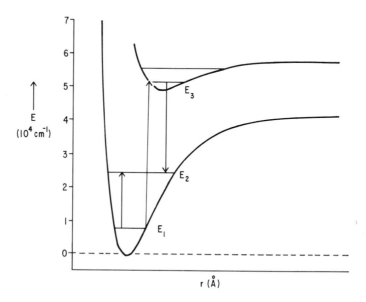

FIG. 5.6.   Transitions giving rise to absorption and emission spectra.

Since each molecule has its own characteristic set of discrete energy levels, it will also have a characteristic set of wavelengths at which it will absorb or emit radiation. It should be noted that all energy changes, or transitions, are not equally probable, so that more light may be absorbed at one wavelength than at another.

In an absorption experiment, the light which passes through the sample is related to the incident light intensity and the sample concentration by the Beer–Lambert–Bouguer equation:

$$I_\lambda = I_0 e^{-\alpha_\lambda c z} \tag{5.3}$$

Many systems may exhibit departures from this simple equation, so that it is usually necessary to construct a calibration curve for the concentration range of interest before using an absorbance measurement as an indication of concentration.

For many emission experiments, the light intensity is directly proportional to the number of molecules of interest in the sample. The constant of proportionality in this case is a sensitive function of all experimental conditions, including the method of excitation, temperature, and other factors in the environment of the excited molecule. Nonlinear effects are generally observed at high concentrations of the emitting species. In principle for a spectroscopic experiment, the wavelengths absorbed or emitted depend on the *identity* and the intensity or absorbance depend on the *amount* of substance.

Since many commercial spectroscopic instruments are available for analytical applications, we shall avoid discussing the details of construction of particular instruments. Neither shall we be concerned with the detailed analyses of molecular spectra which have been used to obtain information about molecular structure and bonding. Since our interest is in the analytical applications of spectroscopy, we shall concern ourselves primarily with sample requirements for various spectrometers and the types of compounds which they can detect.

5.2.1 ULTRAVIOLET-VISIBLE SPECTROMETRY

The wavelength range from 200 nm to about 780 nm (one nanometer, 1 nm, is 10^{-9} m) has long been used for quantitative analysis by absorption spectroscopy. Commercial instruments in which the wavelength is adjusted manually have been workhorses in laboratories where routine analyses of a few substances are being made. A wide range of commercial instruments which automatically plot absorbance as a function of wavelength are also available. These instruments accept solution samples with volumes of a milliliter or so and usually compare the absorbance of a solution containing the substance of interest with the absorbance of a "blank"—a solution which does not contain the substance of interest but which does contain everything else in the sample solution.

The absorption bands of molecules in solution are generally too broad to permit the use of the absorption spectrum as a definitive qualitative method and prior separations must often be performed to eliminate interfering substances. Many organic molecules, particularly those with double bonds, have strong absorption bands at wavelengths shorter than 400 nm. While many simple inorganic ions absorb in all regions of the spectrum, it is common to increase the sensitivity of this method to inorganic ions by forming complex ions. This method is the backbone of the West and Gaeke method for SO_2 and the Saltzman method for NO_2. Absorption spectroscopy affords a simple method for the quantitative analysis of polynuclear hydrocarbons when combined with thin-layer or column chromatography. Detection limits in the range 0.01–0.70 $\mu g/ml$ have been reported for a number of these compounds by Zdrojewski *et al.* (1967).

5.2.2   ATOMIC SPECTROMETRY

In atomic absorption spectroscopy, atoms are produced by aspirating a sample solution into a flame which produces atoms in a low electronic state at the temperatures employed (2000–3000°C). The source of radiation may be a hollow cathode lamp for the element of interest which emits radiation at the same wavelengths at which the atoms absorb. The absorption of radiation by the burner flame at the specific wavelengths of interest within the range 200–800 nm is then a measure of the concentration of that element in the sample. Since atoms do not have vibrational and rotational degrees of freedom, they absorb over relatively narrow wavelength intervals. This means that interferences from other substances are not as common in atomic spectra as in molecular spectra.

The samples are generally prepared in the form of aqueous solutions for this method. Instrument detection limits are reported as ppm (micrograms of substance per milliliter of sample solution) and vary from less than 0.001 ppm to about 100 ppm for different elements. Sample volume requirements fall in the range 1–10 ml/min. Atomic absorption has been applied to the analysis of practically all metallic elements. For high sensitivity, the analysis of each element requires a lamp for that element. A number of fuel–oxidant combinations are used for the flame, depending on the ease with which a given element is volatilized and ionized. Although atomic absorption spectrometry has most often been applied to the analysis of solutions, methods have been devised for reducing and vaporizing metal ions in particulate matter and measuring the absorption directly. Loftin *et al.* (1970) have passed air over carbon rods heated by a radio-frequency generator. The compounds are reduced to atoms by carbon monoxide under these conditions. The atoms are then passed into a hot

TABLE 5.1

COMPARISON OF LIMITS OF DETECTION FOR SEVERAL ELEMENTS[a]

Element	Limit of detection (ppm in solution)		
	Atomic fluorescence	Atomic absorption	Flame emission
Ag	0.0001	0.005	0.02
Au	0.2	0.2	0.5
Be	2.0	0.003	0.2
Bi	0.7	0.05	9.0
Ca	0.02	0.002	0.0001
Cd	0.000001	0.005	0.9
Co	0.1	0.005	0.03
Cr	10.0	0.005	0.005
Cu	0.005	0.005	0.01
Fe	0.25	0.005	0.03
Ga	1.0	0.07	0.01
Ge	10	1.0	0.5
Hf	3.0	15.0	—
Hg	0.1	0.5	20.0
In	0.1	0.05	0.005
Mg	0.008	0.0003	0.004
Mn	0.006	0.002	0.005
Mo	2.0	0.03	0.09
Ni	0.04	0.005	0.03
Pb	0.5	0.03	0.3
Sb	0.4	0.1	1.5
Sc	10	0.1	0.03
Se	0.4	0.5	—
Sr	0.03	0.01	0.0002
Te	0.5	0.3	30.0
Ti	6.0	0.1	0.2
Tl	0.008	0.025	0.002
U	5.0	30.0	30.0
Zn	0.00004	0.002	100.0
Zr	4.0	5.0	3.0

[a] Reprinted from Zacha *et al.*, *Anal. Chem.* **40,** 1733 (1968). Copyright 1968 by the American Chemical Society. Reprinted by permission of the copyright owner.

cell where the concentration is determined by absorption. The sensitivity of this method to lead is 0.16 μg/m^3, well below the levels found in most urban environments. This method would appear to be very useful in those applications where it is necessary to have real-time analysis of the element. A number of commercial atomic absorption units are available. Although atomic absorption has not been widely applied to the analysis of air pollutants, it would appear to have considerable potential in the elemental analysis of particulates.

Much of the instrumentation required for atomic absorption spectroscopy may also be used for atomic emission and atomic fluorescence spectroscopy. In these two methods, atoms are also produced in high-temperature flames. In atomic emission studies, the emission of radiation from atoms which have been thermally excited by the flame is detected. In atomic fluorescence the atoms are excited to high-energy states by radiation; the radiation emitted when the atoms drop to a lower-energy level is detected. The three methods for studying the spectra of atoms are somewhat complementary with respect to limits of detection for various elements, as may be seen in Table 5.1, compiled by Zacha *et al.* (1968). These limits of detection should not be used in an absolute sense since the actual values will depend on burner conditions, the nature of the sources, and other instrumental variables.

One of the older forms of emission spectroscopy does not require the automatic instrumentation common to atomic absorption and fluorescence spectroscopy. Instead, the excited atoms are generated in an arc and the spectra recorded on a photographic plate. This method has been used for several decades in the analysis of metallic elements and although the detection limits are the same order of magnitude as those obtained by atomic absorption and fluorescence, more time and care are required to complete the analysis. While the carbon arc has been widely used for the analysis of relatively abundant elements, a number of alternative methods of creating excited atoms have been developed which increase the sensitivity of the method. Some of these methods have been surveyed by Mitteldorf (1965).

5.2.3 FLUOROMETRY

Fluorometry is a form of emission spectroscopy where light is used as the excitation source.* The emitted radiation may be observed at some

---

* We shall use the single term "fluorometry" to include florimetry, fluorescence spectrometry, spectrofluorimetry, and other similar terms which may be found in literature. Distinctions between these terms need not concern us here.

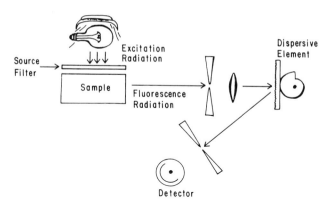

Fig. 5.7.   Basic elements of a fluorescence spectrometer.

angle to the exciting radiation as shown in Fig. 5.7. The sample molecules are raised to excited electronic states by the exciting radiation, which is usually visible or ultraviolet light; the fluorescence is usually observed at wavelengths longer than the source wavelength, generally in the wavelength range 300–800 nm. A variety of commercial instruments are available for making fluorometric analyses. In the simplest devices, resolution of the light is achieved with filters. In somewhat more sophisticated instruments, the excitation readiation is controlled and the fluorescence radiation is analyzed by a dispersive element, such as a grating. In the most elaborate instruments, the excitation wavelength may also be varied continuously. In these latter instruments, the *excitation spectrum* may be obtained by observing changes in emitted intensity (at a fixed wavelength) while changing the excitation wavelength. The advantage of this capability is that an excitation wavelength may be chosen for which the fluorescence is a maximum for the substance of interest. The sample is usually in the form of a solution having a volume of 0.1–1.0 ml.

Many organic molecules exhibit strong fluorescences. Inorganic ions may be studied by incorporating them in organic complexes.* Since a large number of compounds fluoresce and since the fluorescence bands for any one molecule are usually quite broad, it is generally necessary to separate the substance of interest from interfering substances. Chemical separation methods and thin-layer or column chromatography separations are frequently employed. As an example, benzo[a]pyrene may be determined in particulate matter by first dissolving the organic fraction in an appropriate

---

* The fluorescence spectra of atoms in flames may be used to analyze for many metallic elements. This method is discussed briefly in the section on atomic spectrometry.

solvent, separating the organic components with column chromatography and quantitatively measuring the benzo[a]pyrene by fluorometry. Although chromatography may not separate benzo[a]pyrene from benzo[k]-fluoranthene, quantities of both substances may be determined by appropriate fluorometric measurements (Dubois et al., 1967).

The detection limits for some organic compounds are lower for fluorescence measurements than they are for absorption measurements and the two methods tend to complement one another. Thus, although the detection limit for benzo[a]pyrene by absorption is rather high, 0.70 μg/ml, the detection limit for this compound measured by fluorometry is 0.04 μg/ml (Sawicki et al., 1960).

5.2.4 INFRARED SPECTROMETRY

The infrared region of the spectrum extends from the point where light is no longer visible to the human eye, at a wavelength of about 780 nm, to the point where radiation assumes many of the properties of high-frequency radio waves, at wavelengths near 0.5 mm. For most analytical purposes, the infrared is that region from 2.5 to 25 μm. In this type of absorption spectroscopy, the observed transitions are due to changes in the vibrational energy of the lowest electronic state of the molecule.

A variety of commercial instruments are available. They differ primarily in the wavelength range covered and the accuracy with which wavelength and absorbance may be measured. Most instruments accept solution samples or solids dispersed in KBr pellets. Some instruments accept low-temperature sample cells and special, long-path cells for studying low-pressure gases.

Infrared spectroscopy has been very useful to organic chemists in particular because most functional groups, such as the —CHO group in an aldehyde, have a characteristic set of absorption bands which change relatively little from one compound to another. Thus, by examining the infrared spectrum of an unknown compound, one can tell with a fair degree of certainty which functional groups are present. The infrared spectrum of a substance observed in photochemical smog by Stephens et al. (1956) in conjunction with other evidence led to the determination of the structure of PAN.

Liquid-phase spectra may be obtained from neat liquids and from solutions in solvents which are transparent in the wavelength range of interest. NaCl is frequently used for sample cell windows. Carbon tetrachloride and chloroform are used as solvents in addition to many other common substances.

The spectra of solids may be obtained by dispersing the sample in a viscous oil (Nujol) or compressing the solid in a KBr pellet. Nujol and KBr are both transparent over wide ranges of the infrared spectrum. The spectra of solids may also be obtained from thin films deposited on KBr or NaCl windows.

The fact that nitrogen, oxygen, and argon do not absorb in the infrared means that these constituents do not interfere with the analysis of atmospheric gases. Water vapor and carbon dioxide do absorb strongly in the infrared. Many instruments will accommodate a gas cell of about 10 cm in length. For trace constituents, it is necessary to use special cells which achieve very long path lengths with a system of mirrors.

Although infrared spectroscopy is most often applied to the analysis of organic compounds, polyatomic inorganic ions also exhibit characteristic infrared spectra. The relative utility of infrared spectroscopy for inorganic analysis is somewhat diminished by the fact that more sensitive alternatives do exist. (See also nondispersive infrared, p. 150.)

The fluorescence spectra of compounds have also been observed in the infrared portion of the spectrum. Hailey *et al.* (1971) have used laser-induced infrared emission to analyze for a variety of compounds. Special cells must be constructed and the detection limits for such compounds as acetone, chloroform, and ethylether are of the order of 1 ppm.

The principal applications of infrared spectrometry have been to qualitative analysis of unknown compounds; however, many instruments are capable of making reliable quantitative analyses. The use of absorbance as a measure of concentration is often limited by the close proximity of absorption bands of other molecules; however, in favorable cases, reliable concentration versus absorbance curves may be obtained.

5.2.5 NUCLEAR MAGNETIC RESONANCE

Nuclear magnetic resonance spectroscopy measures the relative energies of the nuclear spin states of a molecule in a magnetic field. At the magnetic fields commonly available, the frequency of the radiation used is in the radiofrequency range of the spectrum, generally less than 200 MHz. This method may be applied to the study of any atom having a nuclear spin; most applications at present have been concerned with the ^1H, ^{19}F, and ^{31}P nuclei.

The method is capable of very high resolution and may be used for quantitative and qualitative analysis. As an example, isopropyl benzene, $C_6H_5CH(CH_3)_2$, will exhibit resonances in three areas as shown in Fig. 5.8. The areas under the three groups of curves will be in the ratio of the num-

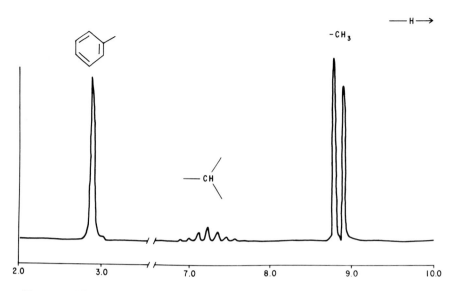

FIG. 5.8.   Nuclear magnetic resonance spectrum of $C_6H_5CH(CH_3)_2$. Horizontal scale is in τ units (ppm).

ber of protons of each type, 5:1:6. The smaller splittings within each group are due to coupling between protons and need not concern us here.

A variety of commercial NMR spectrometers are now available and the method has become a very powerful tool for the organic chemist, primarily in analyzing compounds of unknown structure. While spectra are ordinarily obtained on a milliliter or so of neat liquid or solution, microtubes are available which reduce the sample volume requirement down to about 40 μl. Lundin *et al.* (1966), using computer averaging for a period of one day, have pushed the detection limit down to 0.3 μmole of hydrogen. Brame (1965) suggests a sample limit of 100–400 μg without time averaging. Although the sample requirements for NMR are large enough to preclude its use for routine air monitoring, it remains a useful tool for determining the structures of proton-containing molecules.

5.3 Nuclear Methods

The nuclear method most often used in the analysis of trace constituents is neutron activation analysis. The sample is subjected to a neutron flux which can create radioactive isotopes or metastable nuclei. While these unstable species may decompose in a variety of ways, the gamma ray spectrum of the sample is commonly measured. This spectrum is obtained by allowing the gamma rays to impinge on a detector which has the property of putting out a signal the strength of which is proportional to the energy of the gamma ray. A scaling device measures the energy of each gamma ray and a multiple-channel analyzer records the number of gamma rays emitted within each energy interval. After a short period of time, a record may be obtained of the number of gamma rays as a function of energy. Since the gamma energy depends on the element undergoing decomposition, this record serves as a measure of the variety and quantity of elements present in the sample. Sodium iodide crystals doped with a small amount of thallium emit visible light, the intensity of which is proportional to the gamma ray energy. P-type germanium diodes containing a small amount of lithium produce a voltage pulse proportional to the gamma ray energy. The Ge(Li) detector has significantly more resolution than NaI(Tl), precluding the need for chemical separations in most cases.

The method may be used for the analysis of most elements other than the very lightest nuclei. To a first approximation, this method gives no indication of the chemical state of the element. The amount of sample required depends primarily on the sensitivity of the method to the element of interest. The sample is sealed in a polyethylene vial and irradiated for a fixed length of time. After irradiation, the sample may be counted without further processing if there are no interfering elements present (those which have gamma energies which cannot be resolved from the element of interest by the detector). If interfering elements do exist, further chemical separations are usually required.

To reduce errors due to the chemical processing of the minute samples used in this method, a relatively large amount of nonradioactive carrier of the element of interest, or a small amount of a radioactive tracer of the element, say a beta emitter, may be added to the sample. After the chemical separation, the efficiency of the chemical steps may be determined by conventional chemical means if the nonradioactive carrier is used, or by determining the beta activity if a tracer is used. The analysis may then be completed by measuring the gamma activity.

The detection limits for this method depend on the period and intensity of the neutron activation and on the decay step. It is generally less than 1 μg for most elements. Additional information on the use of this method is given by Guinn and Lukens (1965).

5.4 Mass Spectrometry

Mass spectrometry analyzes a substance by forming ions and then sorting the ions by mass in an electric or magnetic field. The basic experiment is

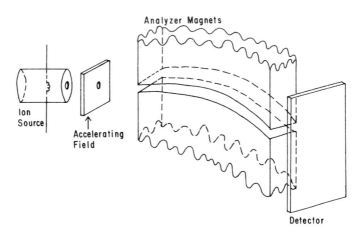

FIG. 5.9. Basic elements of the mass spectrometer. Ions are generated in the ion source, accelerated by an accelerating field passed through the analyzer where they are sorted by mass, using magnets, and finally detected.

entirely different than the spectroscopic experiments we have been discussing. The earliest mass spectrometers were similar in principle to the type diagrammed in Fig. 5.9. Positive ions are produced in the ion source by electron bombardment or an electric discharge. These ions are then given reasonably uniform energies by an electric accelerating field. They next pass through the analyzing section—in this case, a magnetic field at right angles to the plane of the diagram. The moving charges are deflected in the magnetic field by an angle inversely related to the mass of the ion and collected at the detector. The detector may be a photographic plate or an electronic ion detector. Today, there are a number of different types of mass spectrometers available using different methods for ionizing the sample, obtaining mass resolution, and detecting the ions. Although samples are most often introduced as vapors, methods have been developed for obtaining the mass spectra of substances with very high melting points.

In many cases, the parent molecule becomes fragmented during and after the ionization process and a number of daughter ions are produced. The fragmentation patterns which result may be used to identify the substance. The intensity of the mass peaks is a measure of the concentration of the substance in the sample.

The detection limit of mass spectrometers is of the order of one part in 10^7 of the material in the ion source. Thus, in most air quality studies, the air sample must first be concentrated by adsorption, cold trap, or filtration methods in order to reliably measure substances at concentrations of less than 1 ppm.

One of the earliest applications of mass spectrometry to the analysis of air pollutants was the study by Shepherd *et al.* (1951) in which a number of components of Los Angeles County smog were identified. In this study, up to 100 liters of air were passed through a cold trap in order to eliminate oxygen and nitrogen. The sensitivity of the method was estimated to be 10^{-4} ppm of substance in the original air sample. A number of studies since have refined the procedures for collecting samples and obtaining quantitative estimates of the composition of the sample.

Methods have also been developed for the direct introduction of solids into the ion source (Gohlke, 1963). Brown and Vossen (1970) have analyzed for a large number of elements collected on a millipore filter, using a spark source with silver as an internal standard. Arsenic at 0.005 $\mu g/m^3$ and lead at 0.004 $\mu g/m^3$ were reported using a 10.8-m^3 air sample.

Mass spectrometers may also be coupled with other analytical techniques such as the chromatographic methods and various spectroscopic methods. Some of these applications are discussed in the book by Ettre and MacFadden (1969).

5.5 Remote Sensing Applications of Lasers

There are a number of ways of using lasers for the remote sensing of atmospheric constituents. In one basic setup, the laser emits a short pulse of radiation which is returned to the detector as shown in Fig. 5.10. Some of the radiation is also transmitted through the atmosphere. If the outgoing pulse has wavelength λ_0, the back-scattered pulse will contain a component of this same wavelength due to the Rayleigh scattering of the atmospheric gases and the Mie scattering of the aerosols. The returning pulse will also contain components of different wavelengths due to Raman scattering.

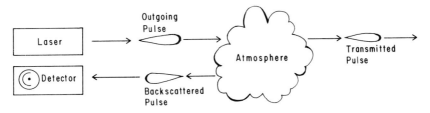

FIG. 5.10. Atmospheric laser experiments. The outgoing pulse is monochromatic. The back-scattered pulse will contain components of this same wavelength due to Rayleigh and Mie scattering and components of other wavelengths due to Raman scattering. The intensity of the transmitted pulse is a measure of total extinction due to scattering and absorption.

The Mie-scattered component may be detected and is a measure of the aerosols present in the scattering volume. Back-scattering by aerosols in the atmosphere has been studied by Barrett and Ben-Dov (1967) and Johnson and Uthe (1971). The Raman back-scattered component can serve to identify the gases present in the scattering volume. Inaba and Kobayasi (1969) have made a preliminary study of the applicability of this method which indicates that the detection limit for CO_2 should be of the order of 0.1 ppm.

Hinkley and Kelley (1971) have considered the use of a laser in a transmissometer in which the intensity of the transmitted pulse would be measured. The use of a tuneable infrared laser would permit the detection of a large number of gases with detection limits of the order of 1 ppb over a 1-km path.

Infrared lasers could also be used to detect infrared radiation emitted from sources with relatively high concentrations, such as plumes. In this application, the emitted radiation is mixed with the laser radiation in a device which produces lower-frequency radiation which may in turn be amplified and detected.

Although laser methods are still in the developmental stage, they are mentioned here because they should be free of interferences for small molecules and should have considerable potential for remote detection.

5.6 Correlation Spectrometry

Correlation spectrometry differs from other types of spectrometry already considered in that the observed spectrum is correlated with the spectrum of a known substance by the instrument. The principles of operation of the correlation spectrometer have been discussed by Davies (1970) and Newcomb and Millan (1970). Some of the essentials of a correlation spectrometer are shown in Fig. 5.11. The refractor plate oscillates at an audio frequency which causes the spectrum of the incident light to jump back and forth on the exit mask which is a replica of the actual spectrum of the gas being analyzed. The photomultiplier tube sees a signal which varies at the frequency at which refractor plates are vibrating. A tuned amplifier and synchronous detector then amplify the output of the photomultiplier tube, producing a dc output. If SO_2 is being measured, the grating is set at a position where there will be minimum interference from NO_2 and other absorbing gases and slowly varied over a small wavelength interval. The result is a slowly varying ac signal the amplitude of which is proportional to the amount of SO_2 in the light path.

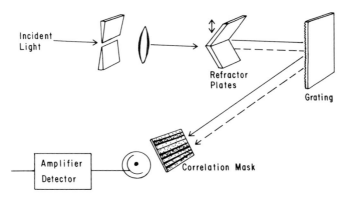

FIG. 5.11.   Correlation spectroscopic apparatus. (Newcomb and Millan, 1970; reproduced by permission of the Institute of Electrical and Electronic Engineers.)

This instrument may be used as a fixed cell monitor in which the sample is pumped through a sample cell illuminated by an artificial light source. It may also be operated as a long-path instrument by simply aligning the instrument with an artificial light source placed up to 1000 m away. No cell is necessary in this arrangement and the instrument determines the concentration of the absorber integrated over the path length, $\int_0^l c\, dl$. The instrument has a detection limit of about 0.03 ppm in a 2.5-m path with a time constant of 100 sec in fixed-path operation. In long-path operation, the detection limit for the integral $\int c\, dl$ is about 2 ppm meter for SO_2 and 1 ppm meter for NO_2 with time constants of the order of 30 sec. The wavelength for SO_2 detection is 280–315 nm and that for NO_2 is 413–450 nm.

The most interesting mode of operation of this instrument is the truly remote mode in which natural radiation, such as the sun or skylight, serves as the source (Moffat and Millan, 1971). In this case, the instrument measures $\int c\, dl$ over the path length between the instrument and the source. This mode of operation has been used in mobile monitoring and would appear to have potential for the monitoring of large areas. Although relatively few atmospheric trace gases absorb in the visible–ultraviolet region of the spectrum, the principle of correlation spectrometry may also be extended to the infrared region of the spectrum.

GENERAL REFERENCES

Bellamy, L. J. (1958). "The Infrared Spectra of Complex Molecules." Wiley, New York.
Christian, G. D., and Feldman, F. J. (1970). "Atomic Absorption Spectroscopy." Wiley, New York.

Conley, R. T. (1966). "Infrared Spectroscopy." Allyn & Bacon, Rockleigh, New Jersey.
Dal Nogare, S., and Juvet, R. S. (1962). "Gas-Liquid Chromatography." Wiley, New York.
Harrison, G. R., Lord, R. C., and Loofbourow, J. R. (1948). "Practical Spectroscopy." Prentice-Hall, Englewood Cliffs, New Jersey.
Heftmann, E. (1961). "Chromatography." Van Nostrand-Reinhold, Princeton, New Jersey.
Keulemans, A. I. M. (1959). "Gas Chromatography." Van Nostrand-Reinhold, Princeton, New Jersey.
Lederer, E., and Lederer, M. (1957). "Chromatography," 2nd ed. Amer. Elsevier, New York.
Littlewood, A. B. (1970). "Gas Chromatography," 2nd ed. Academic Press, New York.
Randerath, K. (1966). "Thin Layer Chromatography." Academic Press, New York.
Stahl, E. (1969). "Thin Layer Chromatography," 2nd ed. Springer Publ., New York.
Udenfriend, S. (1962). "Fluorescence Assay in Biology and Medicine." Academic Press, New York.

REFERENCES

Barrett, E. W., and Ben-Dov, O. (1967). *J. Appl. Meteorol.* **6,** 500.
Brame, E. G. (1965). *Anal. Chem.* **37,** 1183.
Brown, R., and Vossen, P. G. T. (1970). *Anal. Chem.* **42,** 1820.
Darley, E. F., Kettner, K. A., and Stephens, E. R. (1963). *Anal. Chem.* **35,** 589.
Davies, J. H. (1970). *Anal. Chem.* **42** (6), 101A.
Dubois, L., Zdrojewski, A., Baker, C., and Monkman, J. L. (1967). *J. Air Pollut. Contr. Ass.* **17,** 818.
Ettre, L. S., and McFadden, W. H., eds. (1969). "Ancillary Techniques of Gas Chromatography." Wiley, New York.
Fiegl, F. (1958). "Spot Tests in Inorganic Analysis," 5th ed. Amer. Elsevier, New York.
Fiegl, F. (1966). "Spot Tests in Organic Analysis," 7th ed. Amer. Elsevier, New York.
Gohlke, R. S. (1963). *Chem. Ind. (London)* p. 946.
Guinn, V. P., and Lukens, H. R. (1965). *In* "Trace Analysis" (G. H. Morrison, ed.). Wiley (Interscience), New York.
Hailey, D. M., Barnes, H. M., and Robinson, J. W. (1971). *Anal. Chim. Acta.* **56,** 175.
Herzberg, G. (1950). "Spectra of Diatomic Molecules." Van Nostrand-Reinhold, Princeton, New Jersey.
Hinkley, E. D., and Kelley, P. L. (1971). *Science* **171,** 635.
Inaba, H., and Kobayasi, T. (1969). *Nature (London)* **224,** 170.
Johnson, W. B., and Uthe, E. E. (1971). *Atmos. Environ.* **5,** 703.
Loftin, H. P., Christian, C. M., and Robinson, J. W. (1970). *Spectros. Lett.* **3,** 161.
Lovelock, J. E., Maggs, R. J., and Adlard, E. R. (1971). *Anal. Chem.* **43,** 1962.
Lundin, R. E., Elskin, R. H., Flath, R. A., Henderson, N., Mon, T. R., and Teranishi, R. (1966). *Anal. Chem.* **38,** 291.
Mitteldorf, A. J. (1965). *In* "Trace Analysis" (G. H. Morrison, ed.). Wiley (Interscience), New York.
Moffat, A. J., and Millan, M. M. (1971). *Atmos. Environ.* **5,** 677.
Newcomb, G. S., and Millan, M. M. (1970). *IEEE Trans. Geosci. Electron.* **8,** 149.
Sawicki, E., Hauser, T. R., and Stanley, T. W. (1960). *Int. J. Air Pollut.* **2,** 253.

Sawicki, E., Stanley, T. W., Elbert, W. C., and Pfaff, J. D. (1964). *Anal. Chem.* **36,** 497.

Shepherd, M., Rock, S. M., Howard, R., and Stormes, J. (1951). *Anal. Chem.* **23,** 1431.

Stephens, E. R., Scott, W. E., Hanst, P. L., and Doerr, R. C. (1956). *J. Air Pollut. Contr. Ass.* **6,** 159.

Zacha, K. E., Bratzel, M. P., Winefordner, J. D., and Mansfield, J. M. (1968). *Anal. Chem.* **40,** 1733.

Zdrojewski, A., Dubois, L., Moore, G. E., Thomas, R. S., and Monkman, J. L. (1967). *J. Chromatogr.* **28,** 317.

CHAPTER 6

THE ATMOSPHERIC CHEMISTRY
OF SULFUR COMPOUNDS

In this and the following chapters, we shall consider selected aspects of the atmospheric roles of a few classes of substances. No attempt will be made to consider all aspects of, say, the chemistry of sulfur compounds in the atmosphere; rather, some features unique to sulfur compounds will be discussed and a few topics in the area of sulfur chemistry will be introduced in order to illustrate more general points.

6.1 Global Considerations

The principal forms of sulfur in the atmosphere are SO_2 and H_2S in the gas phase and sulfate ion, SO_4^{2-}, in the condensed phases. The SO_4^{2-} ion may be associated with a variety of cations; however, we shall not be concerned with the nature of the cations in much of the following discussion. In humid conditions near large industrial sources of sulfur, the sulfate ion may exist as dilute solutions of sulfuric acid. In the stratosphere, sulfate aerosols have been identified as ammonium sulfate, $(NH_4)_2SO_4$.

Although the man-made, or artificial, sources of atmospheric sulfur may be enumerated relatively easily [see Table 6.1, and SCEP (1970)], the magnitudes of the natural sources of atmospheric sulfur are not at all well understood. Significant natural sources include H_2S formed by bacterial reduction of SO_4^{2-} and decomposition of organic material, sulfate aerosols formed by the foaming and bubbling action of seawater, and volcanoes. Although a reasonably direct measure of the amount of sulfur injected in the form of sea salt aerosols may be obtained, the magnitudes of the biological sources must be obtained indirectly.

As an indication of the difference in the distribution of sulfur sources, Junge (1960) measured the SO_2 and H_2S concentrations under a variety of meteorological conditions in suburban Boston. The H_2S concentration averaged 0.006 ppm and was relatively insensitive to wind direction, suggesting widespread or distant sources. The SO_2 concentration varied from about 0.007 ppm when the wind was coming from a direction in which there were few sources of pollution to about 0.012 ppm under conditions expected to bring about higher ground level concentrations from urban sources.

In the following few paragraphs, we shall discuss some models which have been used in the analysis of the global atmospheric sulfur budget. No attempt will be made to compare the models, because the different authors had access to different types of data. It may be expected that additional models will be constructed in the future as more data become available through increased monitoring activities.

TABLE 6.1

ANTHROPOGENIC SOURCES OF ATMOSPHERIC
SULFUR (1965)[a]

Source	Quantity [Tg(S)/yr]
SO_2, Coal combustion	47
SO_2, Petroleum combustion	10
SO_2, Petroleum refining	3
SO_2, Smelting	7
H_2S, Assorted industries	3
Total	70

[a] Robinson and Robbins (1970); 1 Tg = 1×10^{12} g.

It will be useful in this and in following chapters to have some numbers relating the dimensions and mass of the atmosphere. The surface area figures in Table 6.2 were compiled from available atlases (Rand McNally and Co., 1962). The figures for the mass of the atmosphere were obtained from data in "U.S. Standard Atmosphere, 1962" (U.S. Committee on Extension to the Standard Atmosphere, 1962, 1967). The mass of the atmosphere is given by

$$M = S \int_0^\infty \rho(x) \; dx$$

if we assume that the mass is confined to a shell with a thickness small relative to the radius of the earth. Here, S is the surface area of the earth and $\rho(x)$ is the denisty of the atmosphere as a function of height x. The mass is also given by SP/g, where P is standard atmospheric pressure and g is the gravitational constant. A small amount is subtracted from the mass obtained by either method to account for the continental volume above sea level. (The mean continental elevation was taken to be 750 m.)

One indicator which has been used to measure the distribution of atmospheric sulfur is the amount of sulfur as sulfate in rainwater. Junge and Werby (1958) reported observations made at a large number of stations in the United States which determined Cl^-, Na^+, K^+, and SO_4^{2-} among other ions in rainwater. The Cl^- content of rainwater drops from high values of greater than 1 mg/liter observed along the marine coasts to lows of about 0.1 mg/liter observed in the central plans. The weight ratio SO_4^{2-}/Cl^- ($= 0.140$) is constant in ocean water and, by assuming that all of the Cl^- in rainwater originates from sea spray aerosols and that Cl^- in particles

TABLE 6.2

SELECTED DATA RELATED TO THE DIMENSIONS AND
MASS OF THE ATMOSPHERE

Surface areas (km²)	
Total earth	5.1×10^8
Northern Hemisphere land	1.03×10^8
Northern Hemisphere ocean	1.54×10^8
Southern Hemisphere land	0.46×10^8
Southern Hemisphere ocean	2.10×10^8
Mass of the atmosphere (g)	
Total mass	5.2×10^{21}
Mass of the troposphere (to 11 km)	4.0×10^{21}
Number of moles in the atmosphere	1.8×10^{20}

and rainwater is not lost to the gas phase, it is possible to use Cl^- as a tracer and determine the amount of Na^+, K^+, and SO_4^{2-} originating from sources other than sea spray aerosols. By knowing the distribution of rainfall and the distribution of sulfate concentration in the rain, it is possible to calculate the total amount of excess sulfate deposited. For the earth as a whole, taking a value 2.2 mg/liter for the average excess sulfate in continental rain and 0.5 mg/liter for the average excess in oceanic rain, a value of* 360 Tg per year is obtained for the rate of deposition of excess sulfate from the atmosphere. The emission rate of anthropogenic sulfur (in terms of sulfate) is 110 Tg per year (1943 figures), from which it follows that on a global average, 31% of the excess sulfate in rainwater is a result of man's activities. This also assumes that all sulfur is removed from the atmosphere by precipitation.

Eriksson (1963) based an analysis of the global sulfur budget on the amount of sulfate carried by rivers and observed average atmospheric concentrations, among other factors. The figures are given here as amount of the compounds as sulfur. He assumed that 80 Tg/yr were carried to the seas by rivers. From rainwater analyses, 165 Tg/yr were deposited as sulfate from sources other than sea spray aerosols (65 Tg/yr on land areas and 100 Tg/yr on the oceans). The rate of dry deposition is calculated by assuming that all of the trace constituent in a column of air moving downward at a certain velocity (the deposition velocity) is removed from the atmosphere at the ground. Eriksson assumed a global average concentration of 1 $\mu g/m^3$ for SO_2 with a deposition velocity of 2 cm/sec over land and 0.9 cm/sec over the oceans. Using these figures, 100 Tg/yr are deposited on land areas and 100 Tg/yr are deposited on the oceans. The sedimentation rate in the oceans was assumed to be the same as the rate of weathering of sulfur-containing rocks on land areas, 15 Tg/yr. The net amount of sulfate sulfur transported to the atmosphere as sea spray is taken to be 45 Tg/yr. The sources of sulfur from man's activities were taken as 40 Tg/yr for the industrial output (mostly in the form of SO_2) and 10 Tg/yr for the amount of sulfur added to soils as fertilizer (mostly as sulfates). The flow portion of the diagram is completed by assuming that the total amount of sulfur in the atmosphere is constant. The partitioning of sources of H_2S between continental and maritime areas is obtained by balancing the amount of sulfur entering and leaving these two reservoirs.

The atmospheric burdens for SO_2 and H_2S were calculated by assuming that the mixing ratio was constant throughout the atmosphere and that

---

* The teragram, abbreviated Tg, equal to 1×10^{12} gram, will be used to define quantities of materials on the global scale.

the average ground-level concentrations were 1 $\mu g/m^3$ sulfur in the form SO_2 and 6 $\mu g/m^3$ sulfur in the form H_2S. An estimate of the sulfate burden was made using figures from two sources. Using the data of Junge and Manson (1961), the stratospheric sulfate burden was estimated to be 0.04 Tg. The data of Georgii (1970) were used to obtain the estimates of 1.2 Tg sulfate each for the maritime and continental tropospheres.

A partial flow chart which illustrates only the atmospheric portion of the global system in detail is shown in Fig. 6.1. The residence time, lifetime, or turnover time of the gases in the atmosphere may be calculated from the figures given. This time is taken as a measure of the period a given molecule remains in the atmosphere before being removed or changed into something else. If we look at the figures for H_2S, we see that the atmosphere contains 25 Tg as sulfur. If we could turn off the sources of H_2S without affecting the rate at which H_2S is oxidized to SO_2 (280 Tg/yr), it would take 32 days for the H_2S to disappear.* In general, if R represents the rate at which a sink is taking up a substance and Q is the quantity of that substance in a pool, Q/R is the residence time with respect to that particular substance and sink being considered. For SO_2, we can see that the resi-

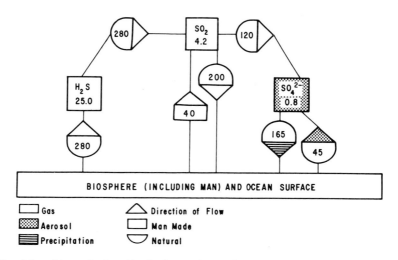

FIG. 6.1.   Atmospheric sulfur budget. The burdens are given in teragrams of sulfur; the flows are expressed in teragrams of sulfur per year.

* If we recognize that reaction rates at which H_2S is converted into something else may be proportional to the concentration of H_2S, then the time we are considering is that required for H_2S to drop to $1/e$ of its initial value (often called the e-folding time). The distinction between these two definitions need not concern us here.

dence time is 13 days with respect to oxidation to SO_4^{2-}, 7.6 days with respect to dry deposition, and 4.8 days with respect to elimination by all means. These figures are global averages and while they may not be related directly to specific rate constants, they may place bounds on the values for the rate constants. More recent estimates of the global sulfur budget using current estimates of anthropogenic emissions have been presented by Robinson and Robbins (1970), Friend (1972), and Kellogg et al. (1972).

6.2 Reactions of Sulfur Compounds

A number of other studies have shown that the residence time of sulfur in the atmosphere as SO_2 is fairly short (ranging from a few hours to a few days, depending on moisture and other conditions). Simultaneous measurements of SO_2 and SO_4^{2-} by Georgii (1970) as a function of altitude suggest that anthropogenic SO_2 predominates over SO_4^{2-} at low altitudes, whereas the SO_2/SO_4^{2-} ratio decreases with altitude. These studies also show that the seasonal variations at low altitudes are greater than those observed at high altitudes, suggesting that the lifetime of sulfur as SO_4^{2-} is longer than the lifetime as SO_2.

Weber (1970) measured the SO_2/CO_2 ratio in power station plumes as a function of time and estimated that the half-life for the disappearance of SO_2 ranged from 20 min to 1 hr, depending on the season. Weber points out that this estimate of the reaction rate holds only for the initial period in the moist plume and very likely may not be applied to the more general atmospheric case where the SO_2 concentration and the relative humidity are lower.

Gerhard and Johnstone (1955) investigated the photochemical oxidation of SO_2 in the laboratory and found a half-life of about 500 hr for the photolysis of SO_2 by intense natural sunlight at concentrations of 1–30 ppm. The reaction was of first order with respect to SO_2 and the rate was unaffected by relative humidity (below saturation), salt nuclei, and NO_2. Junge and Ryan (1958) examined the dark oxidation of SO_2 in water in the presence of $FeCl_2$, which catalyzes the reaction. The possible role which NH_3 may play in the oxidation by increasing the solubility of SO_2 was also discussed. Van den Heuvel and Mason (1963) demonstrated that $(NH_4)_2SO_4$ was formed in water droplets suspended in a stream of air containing SO_2 and NH_3 at the ppm level in the absence of metal ions. These experiments also indicated that the rate-limiting steps occurred in the liquid phase and not in the diffusion processes in the gas phase. Scott and Hobbs (1967) suggested a mechanism for the oxidation of SO_2 whereby the rate-limiting step is the oxidation of the SO_3^{2-} ion which results from the ionization of H_2SO_3, the hydrate of SO_2. The presence of NH_3 plays a significant role in this mechanism by maintaining high concentrations of SO_3^{2-}. Healy *et al.* (1970) have pointed out that in the United Kingdom, enough NH_3 is produced from the decomposition of animal urine to play a significant role in the oxidation of SO_2 when the humidity is near 100%. They suggest that catalysis by transition metal ions might be important at lower relative humidities.

The chemistry of H_2S in the atmosphere has not been studied as extensively. It is generally assumed that H_2S is oxidized to SO_2; however, the conditions which favor the reaction and the mechanism of the reaction are not at all well understood at this time.

6.3 A Mechanism for the Oxidation of SO_2

Examination of the Scott–Hobbs mechanism (Scott and Hobbs, 1967) for the oxidation of SO_2 also brings into play concepts which are required for consideration of the pH of rainwater.

Assume that we have a raindrop in equilibrium with an atmosphere containing NH_3, SO_2, and CO_2, the pressures of which are P_a, P_s, and P_c, respectively. The raindrop will then contain the aqueous forms of these three gases, the concentrations of which will be specified by (NH_4OH), (H_2SO_3), and (H_2CO_3). The concentrations and pressures will be related by Henry's law:

$$P_a = K_{Ha}(NH_4OH) \tag{6.1}$$

$$P_s = K_{Hs}(H_2SO_3) \tag{6.2}$$

$$P_c = K_{Hc}(H_2CO_3) \tag{6.3}$$

The aqueous species being considered all undergo acid–base reactions with water as follows:

$$NH_4OH \rightleftharpoons NH_4^+ + OH^-, \qquad K_a = (NH_4^+)\,(OH^-)/(NH_4OH) \tag{6.4}$$

$$H_2SO_3 \rightleftharpoons H^+ + HSO_3^-, \qquad K_{1s} = (H^+)\,(HSO_3^-)/(H_2SO_3) \tag{6.5}$$

$$HSO_3^- \rightleftharpoons H^+ + SO_3^{2-}, \qquad K_{2s} = (H^+)\,(SO_3^{2-})/(HSO_3^-) \tag{6.6}$$

$$H_2CO_3 \rightleftharpoons H^+ + HCO_3^-, \qquad K_{1c} = (H^+)\,(HCO_3^-)/(H_2CO_3) \tag{6.7}$$

$$HCO_3^- \rightleftharpoons H^+ + CO_3^{2-}, \qquad K_{2c} = (H^+)\,(CO_3^{2-})/(HCO_3^-) \tag{6.8}$$

The water itself will exist in equilibrium with hydrogen and hydroxyl ions:

$$H_2O \rightleftharpoons H^+ + OH^-, \qquad K_w = (H^+)(OH^-) \tag{6.9}$$

The final equation required to solve this problem is the equation of electrical neutrality.

$$(\text{positive charges}) = (\text{negative charges})$$

If we assume that we have an undetermined amount of sulfate ion present and that this ion is a very weak base, then the equation of electrical neutrality takes the following form:

$$(NH_4^+) + (H^+) = (OH^-) + (HSO_3^-) + (HCO_3^-)$$
$$+ 2(SO_3^{2-}) + 2(CO_3^{2-}) + 2(SO_4^{2-}) \tag{6.10}$$

We may solve this equation for (H^+) by substituting Eqs. (6.1)–(6.9) to eliminate the other unknowns. We then obtain

$$a(H^+)^3 - 2(SO_4^{2-})(H^+)^2 - c(H^+) - d = 0 \tag{6.11}$$

with

$$a = 1 + (P_a K_a / K_{Ha} K_w)$$
$$c = K_w + (K_{1s} P_s / K_{Hs}) + (K_{1c} P_c / K_{Hc})$$
$$d = 2(K_{1s} K_{2s} P_s / K_{Hs}) + 2(K_{1c} K_{2c} P_c / K_{Hc})$$

The values of the equilibrium constants used by Scott and Hobbs are

$$K_w = 1.008 \times 10^{-14} \quad \text{(m/liter)}^2 \qquad K_{1s} = 1.27 \times 10^{-2} \quad \text{m/liter}$$
$$K_{Ha} = 1.76 \times 10^{-2} \quad \text{atm (m/liter)}^{-1} \qquad K_{2s} = 6.24 \times 10^{-8} \quad \text{m/liter}$$
$$K_{Hs} = 0.81 \quad \text{atm (m/liter)}^{-1} \qquad K_{1c} = 4.45 \times 10^{-7} \quad \text{m/liter}$$
$$K_{Hc} = 29 \quad \text{atm (m/liter)}^{-1} \qquad K_{2c} = 4.68 \times 10^{-11} \quad \text{m/liter}$$
$$K_a = 1.774 \times 10^{-5} \quad \text{m/liter}$$

Equation (6.11) may be used to calculate (H^+) if we know the partial pressures of the three gases and (SO_4^{2-}). It is assumed that the droplet has not absorbed any other acidic or basic substances, such as HCl or alkaline dusts (carbonates). For final consideration of this mechanism, we need a rate constant expression for the oxidation of SO_3^{2-}. This is given by

$$d(SO_4^{2-})/dt = k(SO_3^{2-}) \tag{6.12}$$

We also need an equation relating (H^+) to (SO_3^{2-}); this may be obtained from Eqs. (6.5) and (6.6) as

$$(SO_3^{2-}) = K_{1s}K_{2s}P_s/K_{Hs}(H^+)^2 \qquad (6.13)$$

Assuming a value of 0.1 min^{-1} for the rate constant k, and assuming values for the pressures of the three gases, we can calculate the rate of increase of (SO_4^{2-}) in the raindrop. While these equations may be solved for certain limiting cases, a computer must be used to obtain more general solutions.

6.4 Analytical Methods

6.4.1 WEST-GAEKE METHOD FOR SO$_2$

The most widely used chemical method for the analysis of SO$_2$ is the West–Gaeke method. The outline given here is taken from descriptions given by West and Gaeke (1956) and Nauman *et al.* (1960) and in "Selected Methods for the Measurement of Air Pollutants" (U. S. Dept. of Health, Education, and Welfare, 1965).

Sulfur dioxide is absorbed from the atmosphere by a sodium tetrachloromercurate(II), Na$_2$HgCl$_4$, solution, which stabilizes the SO$_2$ as sodium dichlorosulfitomercurate(II), preventing reentrainment of the SO$_2$. Sulfur dioxide may be stored in this form for periods of up to a week before completing the analysis. The analysis is completed by adding acidic pararosaniline and formaldehyde to the complex mercurate ion, which removes the SO$_2$ unit, forming the reddish pararosaniline methyl sulfonic acid, the concentration of which is determined spectrophotometrically. The detection limit for this method is about 0.005 ppm and other acidic and basic substances do not interfere, as they do with conductimetric and hydrogen peroxide methods.

Reagents

0.1 M Na$_2$HgCl$_4$.   Dissolve 0.1 mole of HgCl$_2$ and 0.2 mole of NaCl in distilled water. Dilute to one liter.

Pararosaniline 0.04%.   Make up a stock 0.2% pararosaniline solution by dissolving 0.2 g of pararosaniline hydrogen chloride in 100 ml of water. Filter after 48 hr. Add 20 ml of the 0.2% pararosaniline to 6 ml of concentrated HCl, allow to stand 5 min, then dilute to 100 ml with distilled water. Both pararosaniline solutions are light-sensitive and should be stored in dark bottles in a refrigerator.

Formaldehyde 0.2%.   Dilute 5 ml of 40% formaldehyde to one liter with distilled water. Prepare weekly.

Procedure

Set up a sampling train consisting of an air pump, flowmeter, thermometer, and a midget impinger containing exactly 10 ml of the absorbing solution. (Two impingers may be used in series to test the collection efficiency.) The pressure should also be measured at the flowmeter. The various components should be connected with glass or Teflon tubing or something which does not react with SO_2. The impinger should be protected from light. Draw air through the system at a rate of 0.2–2.5 liters/min until the desired amount of SO_2 is obtained. (The greatest relative precision for an absorption experiment will be obtained when the photometer reads at about the middle of the dial—actually, at 37% transmission.)

After sampling, filter the solution if any precipitate is present (due to re-duced sulfur compounds in the air) and adjust the volume to 10 ml. Add 1.0 ml of the acidic pararosaniline and 1.0 ml of the formaldehyde solution, mix, and allow to stand for 20 min. A blank should be prepared by treating 10 ml of Na_2HgCl_4 to all steps except exposure to the air stream. The absorbance is read at 560 nm using the blank as a reference.

Calibration curves of absorbance versus SO_2 concentration may be obtained directly by using an SO_2 permeation tube or indirectly by using a standard metabisulfite solution. The metabisulfite solution may in turn be standardized with a standard iodine solution.

Interference by ozone and nitrogen oxides has been shown by Terraglio and Manganelli (1962) to be important only when the mass concentrations of these substances are the same magnitude as that of sulfur dioxide. Although the dichlorosulfitomercurate ion is reasonably stable for periods of up to a week, Scaringelli et al. (1970) have increased the stability of SO_2 with respect to oxidation by adding EDTA.

6.4.2 HYDROGEN PEROXIDE METHOD

This method requires less equipment than the West–Gaeke method. General references for this method are "Selected Methods for the Measurement of Air Pollutants" (U. S. Department of Health, Education, and Welfare, 1965) and Greenburg and Jacobs (1956). The precision which may be attained by the two methods are comparable. Sulfur dioxide is absorbed in a buffered H_2O_2 in which it is oxidized to sulfuric acid. The amount of sulfuric acid formed is determined by titration with a standard base.

Reagents

0.03 N H_2O_2 adjusted to pH 5.   Dilute 3.4 ml of 30% H_2O_2 to two liters. Determine the amount of acid required to reach pH 5 by adding a few drops of mixed indicator to a 75-ml aliquot of the 0.03 N H_2O_2 and then adding dropwise 0.002 N HCl or HNO_3 until the indicator turns pink. Then add the amount of acid calculated to bring the remaining solution to pH 5.

Mixed indicator.   Dissolve 0.06 g of bromcresol green and 0.04 g of methyl red in 100 ml of methanol. The indicator equivalence point is green for the $NaOH–H_2SO_4$ titration.

Standard H_2SO_4, 0.002 N.   Prepare by dilution and standardize with an appropriate acidimetric standard.

Standard NaOH, 0.002 N.   Prepare using carbon dioxide-free water and protect from absorption of CO_2 from the atmosphere. Standardize with the standard sulfuric acid.

Procedure

Place 75 ml of the H_2O_2 solution in an all-glass impinger with a fritted bubbler. After sampling for an appropriate length of time at 2–3 liters/min, add three drops of the indicator to the solution and titrate with the standard NaOH. A 75-ml reagent blank which has not been exposed to SO_2 is also titrated. The amount of NaOH required to titrate the blank is subtracted from that required to titrate the sample. (The blank titer should be less than 0.1 ml.)

Any acidic or basic substances absorbed by the solution will interfere with this determination of SO_2. In calculating the number of moles of SO_2 in the sample, it should be remembered that two moles of sodium hydroxide are required to react with each mole of sulfur dioxide.

6.4.3   THORIN METHOD FOR SULFATE

Sulfur as SO_4^{2-} may be determined by using Thorin as an indicator (Fritz and Yamamura, 1955). Thorin forms a pink complex with Ba^{2+} and the analysis is performed titrimetrically, adding $Ba(ClO_4)_2$ to the sulfate unknown in isopropanol–water solution with Thorin. When all of the free sulfate has been precipitated by the barium ions, excess Ba^{2+} forms the pink complex with Thorin. Fielder and Morgan (1960) have shown that this method may be used to analyze for sulfur in the +6 oxidation state in the presence of large amounts of SO_2.

6.4.4 METHYLENE BLUE METHOD FOR REDUCED SULFUR

Reduced sulfur, as H_2S and mercaptans, may be analyzed by the colorimetric methylene blue method (Jacobs *et al.*, 1957). The sulfur compounds are absorbed by alkaline cadmium hydroxide, which prevents oxidation or reentrainment of the sulfur. The intensely colored methylene blue is formed upon the addition of p-aminodimethylaniline and ferric chloride. The absorbance is measured at 670 nm. While no detection limits are given for this substance, the molar absorptivity (ϵ = absorbance/c_S) is about 35,000 per mole, and Jacobs reports H_2S concentrations down to the 0.2-ppb level.

6.4.5 AUTOMATIC INSTRUMENTATION

A recent development in automatic instrumentation is the flame photometric method for sulfur. When sulfur compounds are burned in a reducing

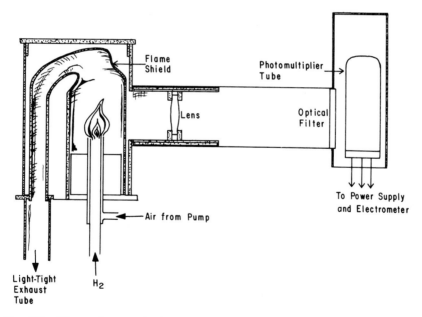

FIG. 6.2. Flame photometric detector for sulfur compounds. The emission of the S_2 molecule, formed in the flame, is a measure of the concentration of sulfur compounds in the atmosphere. Reprinted from Crider, *Anal. Chem.* **37,** 1770 (1965). Copyright 1965 by the American Chemical Society. Reprinted by permission of the copyright owner.

hydrogen/air flame, the emission of the S_2 molecule at 394 nm may be used as a measure of sulfur concentration (Brody and Chaney, 1966). This detector has been used as a specific sulfur detector on gas chromatographs and Stevens et al. (1969) have shown that this detector may be used down to the ppb level for sulfur compounds in air without preconcentration. Figure 6.2 illustrates the essential features of this detector. A total sulfur detector such as this could be used in conjunction with selective scrubbers to measure sulfur oxides or reduced sulfur individually (Adams et al., 1968).

Automatic instrumentation is also available for measuring sulfur compound concentrations by conductimetric, coulometric, and colorimetric methods. The performance characteristics of these instruments have been thoroughly examined by Rodes et al. (1969) and Palmer et al. (1969).

PROBLEMS

1. (a) Calculate the pH of a raindrop in equilibrium with 3.2×10^{-4} atm of CO_2.   (pH = 5.66)

   (b) Calculate the pH of a raindrop in equilibrium with 3.2×10^{-4} atm of CO_2 and 5×10^{-7} atm SO_2. (Neglect the oxidation of SO_2 for this problem.) Is the SO_2 or CO_2 more important in determining the pH in this example?   (pH = 4.05)

2. The sulfate collected in 9.8 m³ of air at 25°C and 1 atm was analyzed by the Thorin method. A total of 2.10 ml of standardized $Ba(ClO_4)_2$ were required to precipitate the SO_4^{2-} in the sample. The $Ba(ClO_4)_2$ was standardized by titration against a standard sulfuric acid solution. A total of 21.2 ml of the $Ba(ClO_4)_2$ were required to react with the sulfate in a 20.0-ml sample of the standard acid solution. The sulfuric acid was in turn standardized by titrating 18.6 ml of the acid with 10.0 ml of 0.0220 N NaOH. Assume that the uncertainties in the volumes are 0.05 ml. Calculate the molarities of the $Ba(ClO_4)_2$ and H_2SO_4, the concentration of SO_4^{2-} in $\mu g/m^3$, and the associated uncertainties.

3. In an analysis of a stack sample for SO_2 by the hydrogen peroxide method, 24 liters of air at 27°C and 750 Torr were sampled. A total of 9.64 ml of 0.00184 N NaOH were required to neutralize the sulfuric acid formed. The blank titer was determined to be 0.06 ml. Calculate the concentration of SO_2 in ppm. What information is required to calculate the stack concentration in $\mu g/m^3$?

4. Calculate the concentration of SO_2 emitted during the combustion of coal containing 65% carbon, 7% hydrogen, 0.8% sulfur, and 4% water,

with the remaining fraction being noncombustible and nonvolatile. Assume that a 20% excess of air is used and that all of the sulfur is emitted as SO_2. Will the stack concentration of SO_2 depend on the rate of fuel consumption? Will the concentration of SO_2 several miles downwind depend on this rate?

5. (a) In a calibration of the West–Gaeke method for SO_2, the SO_2 emitted by a permeation tube was diluted and absorbed in 10 ml of solution in a bubbler. The absorbance was then read against a blank with a spectrophotometer. In the first run, the air stream containing the SO_2 passed through the bubbler for 2.5 min and the absorbance was 0.14. In the second run, the SO_2 was collected for 17.5 min and the absorbance was 0.97. The rate of loss of SO_2 by the permeation tube was known to be 9.75×10^{-7} g/min at the temperature used. Prepare a calibration plot of absorbance as a function of the amount of SO_2.

(b) In an application of the system calibrated above, the SO_2 in 5 ft^3, measured at 30°C and 730 Torr, gave an absorbance of 0.52. Calculate the concentration in ppm and $\mu g/m^3$.

6. Show that the initial rate of formation of sulfate [the rate when $(SO_4^{2-}) = 0$] in the Scott–Hobbs mechanism is given by $d(SO_4^{2-})/dt = K'P_a$ if P_s is fairly large (a few ppm) and P_a is small (less than 1 ppm). Evaluate K' $(K' = kK_{2s}K_a/K_{Ha}K_w)$.

7. Show that the rate of formation of sulfate is given by $d(SO_4^{2-}/dt = K''/(SO_4^{2-})^2$ when the sulfate concentration is quite large (low pH). Evaluate K'' $(K'' = kK_{1s}K_{2s}P_sP_a{}^2K_a{}^2/4K_{Hs}K_{Ha}{}^2K_w{}^2)$.

REFERENCES

Adams, D. F., Bamesberger, W. L., and Robertson, T. J. (1968). *J. Air Pollut. Contr. Ass.* **18,** 145.
Brody, S. S., and Chaney, J. E. (1966). *J. Gas Chromatogr.* **4,** 42.
Crider, W. L. (1965). *Anal. Chem.* **37,** 1770.
Eriksson, E. (1963). *J. Geophys. Res.* **68,** 4001.
Fielder, R. S., and Morgan, C. H. (1960). *Anal. Chim. Acta* **23,** 538.
Friend, J. P. (1972). *Science* **175,** 1278.
Fritz, J. S., and Yamamura, S. S. (1955). *Anal. Chem.* **27,** 1461.
Georgii, H. W. (1970). *J. Geophys. Res.* **75,** 2365.
Gerhard, E. R., and Johnstone, H. F. (1955). *Ind. Eng. Chem.* **47,** 972.
Greenberg, L., and Jacobs, M. B. (1956). *Ind. Eng. Chem.* **48,** 1517.
Healy, T. V., McKay, H. A. C., Pilbeam, A., and Scargill, D. (1970). *J. Geophys. Res.* **75,** 2317.
Jacobs, M. B., Braverman, M. M., and Hochheiser, S. (1957). *Anal. Chem.* **29,** 1349.
Junge, C. E. (1960). *J. Geophys. Res.* **65,** 227.
Junge, C. E., and Manson, J. E. (1961). *J. Geophys. Res.* **66,** 2163.
Junge, C. E., and Ryan, T. G. (1958). *Quart. J. Roy. Meteorol. Soc.* **84,** 46.

Junge, C. E., and Werby, R. T. (1958). *J. Meteorol.* **15,** 417.

Kellogg, W. W., Cadle, R. D., Allen, E. R., Lazrus, A. L., and Martell, E. A. (1972). *Science* **175,** 587.

Nauman, R. V., West, P. W., Tron, F., and Gaeke, G. C. (1960). *Anal. Chem.* **32,** 1307.

Palmer, H. F., Rodes, C. E., and Nelson, C. J. (1969). *J. Air Pollut. Contr. Ass.* **19,** 778.

Rand McNally and Co. (1962). "International World Atlas." Rand McNally and Co., Chicago, Illinois.

Robinson, E., and Robbins, R. C. (1970). *In* "Global Effects of Environmental Pollution" (S. F. Singer, ed.). Springer-Verlag, Berlin and New York.

Rodes, C. E., Palmer, H. F., Elfers, L. A., and Norris, C. H. (1969). *J. Air Pollut. Contr. Ass.* **19,** 575.

Scaringelli, F. P., Elfers, L., Norris, D., and Hochheiser, S. (1970). *Anal. Chem.* **42,** 1818.

SCEP, "Man's Impact on the Global Environment" (C. L. Wilson, ed.). MIT Press, Cambridge, Massachusetts.

Scott, W. D., and Hobbs, P. V. (1967). *J. Atmos. Sci.* **24,** 54.

Stevens, R. K., O'Keeffe, A. E., and Ortman, G. C. (1969). *Environ. Sci. Technol.* **3,** 652.

Terraglio, F. P., and Manganelli, R. M. (1962). *Anal. Chem.* **34,** 675.

U.S. Comm. on Extension to the Standard Atmosphere (1962). "U.S. Standard Atmosphere, 1962." U.S. Govt. Printing Office, Washington, D.C.

U.S. Comm. on Extension to the Standard Atmosphere (1967). "U.S. Standard Atmosphere, Supplements, 1966." U.S. Govt. Printing Office, Washington, D.C.

U.S. Dept. Health, Education, and Welfare (1965). Selected methods for the measurement of air pollutants. Publ. 999-AP-11. Publ. Health Serv., Cincinnati, Ohio.

Van den Heuvel, A. P., and Mason, B. J. (1963). *Quart. J. Roy. Meteorol. Soc.* **89,** 271.

Weber, E. (1970). *J. Geophys. Res.* **75,** 2909.

West, P. W., and Gaeke, G. C. (1956). *Anal. Chem.* **28,** 1816.

CHAPTER 7

NITROGEN COMPOUNDS AND OZONE

The five principal nitrogen-containing gases in the atmosphere are nitrogen (N_2), ammonia (NH_3), nitrous oxide (N_2O), nitric oxide (NO), and nitrogen dioxide (NO_2). Other oxides, such as NO_3, N_2O_3, N_2O_4, and N_2O_5, may be important as reaction intermediates; however, their formation is not favored at the low partial pressures generally observed for the simple oxides (see problems at end of chapter). Nitrogen is found in the condensed phase as ammonium ion (NH_4^+) and nitrate ion (NO_3^-). Significant concentrations of organic nitrates have also been observed in urban atmospheres. The global nitrogen budget has been considered by a number of authors, including Robinson and Robbins (1970) and Delwiche (1970). While we do not have the space here to consider all aspects of this problem, we shall attempt to describe in part the participation by the more important nitrogen-containing trace gases in this cycle.

7.1 Reactions of Nitrogen Compounds

Nitrogen gas (N_2) has a large bond energy and is generally inert with respect to ordinary tropospheric processes. Although some bacteria are capable of producing nitrogen compounds from nitrogen, the fraction of the total global nitrogen pool involved in such reactions in a given year is very small. If we assume the mass of the atmosphere is 5.2×10^9 Tg with an average molecular weight of 28.96 g/mole and that the atmosphere contains 78% N_2 by volume, the total amount of nitrogen in the atmosphere is 3.9×10^9 Tg.

Ammonia is widely distributed in the atmosphere both as NH_3 in the gas phase and as NH_4^+ in aerosols, clouds, and rainwater. The sources of ammonia and its fate in the atmosphere are not well understood. The oceans contain organisms capable of producing ammonia from nitrates and may act as a source. In continental areas, ammonia is also formed by the bacterial deamination of protein and by fixation of oxides of nitrogen in the soil. As an example of a particular source, Healy et al. (1970) have estimated that animal urine alone may account for 85×10^9 g/yr of ammonia in the United Kingdom, about 85% of the ammonia produced in this area. Ammonia gas is a weak base and therefore its solubility in rainwater as well as its release from soils or ground water are sensitive functions of pH.

Nitrous oxide is a comparatively inert compound with respect to tropospheric chemical processes and is distributed uniformly throughout the troposphere. Arnold (1954) demonstrated that many soils contained bacteria capable of producing N_2O from either NH_4^+ or NO_3^- and concluded that soils are a major source of this gas. LaHue et al. (1970) observed concentrations at a number of sites in Panama closely grouped around 0.26 ppm; no statistically significant seasonal trends were observed. Schütz et al. (1970) examined the seasonal dependence at a continental site and found that the nitrous oxide concentration was remarkably independent of wind direction. Their data also suggested that the mixing ratio was independent of height in the troposphere. The nitrous oxide cycle appears to have an interface with the cycles of other nitrogen compounds at the surface of the earth, where most of it is produced and decomposed, and near the tropopause, which acts as a sink. At higher altitudes, the ultraviolet light intensity increases and at wavelengths shorter than 340 nm, nitrous oxide is photolyzed to form N_2 and O, or NO and N, depending on the energy of the photon absorbed. An average concentration of 0.25 ppm for N_2O yields 2000 Tg for the amount of N_2O in the atmosphere. Bates and Hays (1967) estimated that 29.4 Tg/yr could be decomposed in the stratosphere, producing, among other things, 4.0 Tg/yr of NO. Arnold (1954) estimated that the soil was capable of producing 900 Tg/yr of N_2O, although this figure was based on laboratory experiments with selected soils.

The atmospheric chemistry of nitrogen dioxide, nitric oxide, and ozone is usually discussed within the context of the photochemical smog common in those urban areas that have: (1) high automobile traffic densities, (2) frequent periods of atmospheric stability or topographic traps, and (3) sunlight. It appears, however, that natural sources of nitrogen dioxide and nitric oxide far outweigh anthropogenic sources on a global basis (Robinson and Robbins, 1970; Altshuller, 1958). Robinson and Robbins assumed an average concentration of 4 ppb for NO_2 over land areas between 65°N and 65°S latitude and 0.5 ppb for the rest of the earth's surface. The corresponding values for NO were taken to be 2 ppb and 0.2 ppb. The amounts of these gases in the atmosphere are then 11.8 Tg (NO_2) and 4.2 Tg (NO). The relatively high concentrations of nitrogen oxides in urban areas and the presumed relationship to many of the symptoms of photochemical smog has intensified the study of reactions of the nitrogen oxides in atmospheres containing other pollutants. The concentrations of nitrogen dioxide, nitric oxide, and ozone are strongly linked by a sequence of reactions which have been discussed in detail by Leighton (1961), Schuck and Stephens (1969), and Altshuller and Bufalini (1971). While the more interested reader is encouraged to refer to these reviews on the subject for more details, we shall discuss a few aspects of the photochemistry of these compounds. Three of the more important reactions in daylight conditions linking NO_2, NO, and O_3 at the concentrations generally observed in the atmosphere are

$$NO_2 \rightarrow NO + O \qquad \phi k_a = 0\text{--}25 \quad hr^{-1} \qquad\qquad I$$

$$O + O_2 + M \rightarrow O_3 + M \qquad k_3 = 8.9 \times 10^{-4} \quad ppm^{-2} hr^{-1} \qquad II$$

$$O_3 + NO \rightarrow NO_2 + O_2 \qquad k_1 = 1320 \quad ppm^{-1} hr^{-1} \qquad\qquad III$$

The rate constants are given for 25°C and assume that the total pressure is 1 atm. The rate constants used are those given by Schuck and Stephens (1969) and Clyne et al. (1964). The value of ϕk_a will depend on the amount of sunlight present; the maximum value given is typical for a sunny day at the latitude of Los Angeles. The factor ϕ defines the efficiency of the primary photochemical step and is a function of wavelength. This factor is negligibly small for $\lambda > 440$ nm, rises to about 0.4 at 400 nm, and is greater than 0.8 for $\lambda < 380$ nm (Schuck and Stephens, 1969). These reactions alone do not account for the observed time dependences of O_3, NO, and NO_2 in photochemical smog; however, these reactions are faster than other reactions involving these constitutents. For this reason, they are felt to be important in establishing a relationship between the concentrations of O_3, NO, and NO_2 in sunlight.

For the three reactions given, the rate of production of oxygen atoms and ozone will be given by

$$d(O)/dt = \phi k_a(NO_2) - k_3(O)(O_2)(M) \tag{7.1}$$

$$d(O_3)/dt = k_3(O)(O_2)(M) - k_1(O_3)(NO) \tag{7.2}$$

If we assume that a dynamic equilibrium or steady state is reached for the oxygen atom concentration, then $d(O)/dt = 0$ and we obtain

$$d(O_3)/dt = \phi k_a(NO_2) - k_1(NO)(O_3) \tag{7.3}$$

The two terms on the right-hand side of Eq. (7.3) are generally much larger than values of $d(O_3)/dt$ observed in the atmosphere. Thus, to some degree of approximation, we may set $d(O_3)/dt$ equal to zero and obtain the following relationship:

$$(NO)(O_3)/(NO_2) = \phi k_a/k_1 \tag{7.4}$$

This equation describes a photochemical equilibrium between these three reactants. The "equilibrium" constant $\phi k_a/k_1$ depends on the amount of radiation present. It should be noted that this relationship should be a valid one, irrespective of what other reactions may occur to remove these substances, as long as the other reactions are much slower than the three reactions given. We may obtain a measure of the time constant for the relaxation to equilibrium for this system if we imagine an experiment whereby we suddenly inject a small amount of nitric oxide into a system containing NO, NO_2, and O_3 at photochemical equilibrium. It may be shown that the restoration of the disturbed system to equilibrium will be a first-order reaction with a half-life of about 16 sec if we assume initial concentrations of 0.05 ppm for NO and O_3 and a value 21 hr^{-1} for ϕk_a. The relaxation time is a measure of the time required for this equilibrium to adjust to changing conditions—light intensity, concentration of NO, NO_2, or O_3. The shortness of the time suggests that the relationship expressed by Eq. (7.4) should be able to adjust continuously to most changes expected in the atmosphere.

The principal sources of nitrogen oxides of concern in urban areas are combustion processes. The equilibrium $N_2 + O_2 \rightleftharpoons 2NO$, which lies far to the left at 25°C, shifts toward the right at the temperatures present in combustion chambers. If the N_2–O_2–NO mixture is cooled rapidly, the reverse reaction, $2NO \rightarrow N_2 + O_2$, is quenched and the return to 25°C equilibrium values proceeds very slowly. Figure 7.1 shows the equilibrium pressures of NO and NO_2 in an atmosphere of 0.035 atm O_2 and 0.78 atm N_2 at various temperatures. This is the amount of O_2 and N_2 expected to be present if a 20% excess of O_2 is added to the fuel. The actual concentrations of nitrogen oxides will deviate from the values shown in the figure

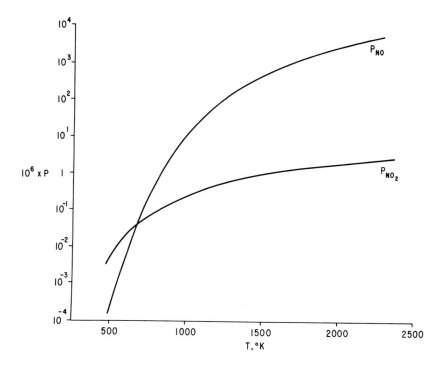

FIG. 7.1.   Equilibrium pressures of NO and NO₂ in a mixture of 0.035 atm O₂ and 0.78 atm N₂.

because the high-temperature equilibrium may not be attained before the reaction products are cooled and because some back reaction may occur during the cooling. The production of NO may be reduced by allowing the gases to cool more slowly.

Anthropogenic nitrogen oxides enter the atmosphere mostly as nitric oxide; however, the conversion of NO to NO₂ takes place within a few hours in urban environments. An estimated 48 Tg/yr of NO and NO₂ (computed as NO₂) are released to the atmosphere by man's activities. Both NO and NO₂ are also produced by bacterial reduction of nitrates. Because it is readily oxidized to NO₂, NO rarely reaches levels which are toxic to man. Highly toxic concentrations of NO₂ (of the order of several hundred ppm) have been encountered in freshly filled silos with poor ventilation (Lowry and Schuman, 1956). The magnitudes of the natural sources of NO and NO₂ will be considered later. A typical daily pattern observed in photochemical smog is shown in Fig. 7.2. In this figure, we can see the rise in

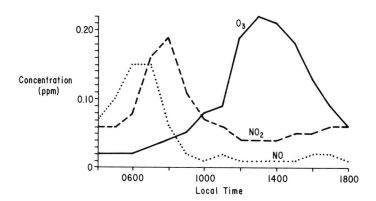

FIG. 7.2. Averaged daily patterns for the time dependence of NO, NO₂, and O₃ (Leighton, 1961; reproduced by permission of Academic Press.)

NO concentration with morning traffic, followed by oxidation of the nitric oxide to nitrogen dioxide, which in turn is followed by an increase in ozone concentration and the appearance of organic nitrates and oxygenated hydrocarbons shortly after noon. The mechanism for the oxidation of nitric oxide is not well understood. It should be noted that this oxidation is not accounted for by the reactions considered earlier in the chapter, but presumably involves the reactions of these compounds with other compounds present in the polluted atmosphere. The "other" compounds probably include organic trace constituents, particularly unsaturated hydrocarbons, and carbon monoxide. At least three forms of elemental oxygen undergo fast reactions with unsaturated hydrocarbons and have the potential of producing reactive intermediates. Ozone and atomic oxygen are very reactive (Altshuller and Bufalini, 1965), and recently a role has been attributed to excited oxygen molecules [see, for example, Coomber and Pitts (1970)]. Oxygen molecules in the excited $^1\Delta$ state may be produced by a variety of energy-transfer processes and could play a significant role if present in large enough concentrations.

The details of the mechanism for the oxidation of NO to NO₂ in the atmosphere are not well understood. The key step may involve a reaction with a free radical produced by the reactions of atomic oxygen, ozone, or excited oxygen molecules with hydrocarbons [see, for example, Friedlander and Seinfeld (1969)]:

$$R + NO \rightarrow R' + NO_2$$

A mechanism involving CO has been proposed by Westberg *et al.* (1971)

in which the three significant reactions are

$$OH + CO \rightarrow CO_2 + H$$

$$H + O_2 + M \rightarrow HO_2 + M$$

$$HO_2 + NO \rightarrow OH + NO_2$$

These are all free-radical chain reactions in which one free radical is consumed and another produced by each step. The overall result is the oxidation of NO and CO:

$$O_2 + CO + NO \rightarrow NO_2 + CO_2$$

Although there is a great deal of empirical evidence concerning the effect of CO, SO_2, hydrocarbons, and other substances on the rate of oxidation of NO and the rate of ozone production, it is not possible at this time to present a detailed mechanism for the formation of photochemical smog.

One of the principal nitrogen-containing organic compounds observed in polluted atmospheres is peroxyacetyl nitrate (PAN):

$$\begin{array}{c} O \\ \| \\ CH_3C-O-O-NO_2 \end{array}$$

This compound does not appear to be formed from hydrocarbons and nitrogen oxides by a few simple chemical reactions such as appears to be the case for the conversion of sulfur dioxide to sulfate ion. PAN and other peroxyacyl nitrates are not particularly stable compounds and very likely undergo further reactions. PAN takes part in a dark reaction with nitric oxide (Schuck and Stephens, 1969):

$$NO + RC(O)-O-O-NO_2 \rightarrow 2NO_2 + RCO_2$$

and decomposes in the presence of base in the condensed phase (Stephens, 1967):

$$OH^- + RC(O)-O-O-NO_2 \rightarrow RCOOH + O_2 + NO_2^-$$

The final fate of the nitrogen emitted as NO is not well understood. Nitric acid has been observed in some laboratory studies, but may result from wall effects.

A number of the symptoms of photochemical smog have been attributed, in part, to the presence of ozone. In addition to its appearance as a secondary product of photochemical smog, ozone is formed naturally in the stratosphere by a series of photochemical reactions. The initiating step in the natural production of O_3 is the photolysis of O_2, producing oxygen atoms. These atoms react with O_2, forming O_3 via the reaction discussed earlier. The absorption of solar radiation in the interval 200–300 nm, which results in the heating of the stratosphere, is largely due to the presence of O_3 [see, for example, Craig (1965)]. This absorption also prevents significant amounts of ultraviolet radiation from reaching the lower altitudes. Interest in the effect of high-altitude aircraft flights on the stratosphere has stimulated recent considerations of the reactions of ozone in the stratosphere (Harrison, 1970; Johnston, 1971).

Natural ozone reaches ground level by advection and dispersion processes, where it is destroyed. Regener (1957) has estimated the ozone flux to be about 1×10^{11} molecules/cm^2-sec from measurements of the concentration gradient near the ground.

7.2 Global Aspects

The atmospheric nitrogen cycle will be constructed using data presented by Robinson and Robbins (1970), some of which are presented elsewhere in this chapter. We shall express the quantities in terms of teragrams of nitrogen. The N_2O portion of the cycle has already been considered. The value 600 Tg/yr for the atmospheric production of N_2O was obtained from the estimate of Arnold (1954). Although this figure is highly uncertain, the N_2O portion of the nitrogen cycle does not appear to interact with other portions of the atmospheric cycle to a significant extent.

The NH_3/NH_4^+ cycle may be constructed on the basis of the deposition rates of NH_4^+ and NH_3. Robinson and Robbins assume that the dry deposition of NH_4^+ is 25% of the deposition by rainfall. The deposition rate for NH_3 is obtained by assuming that the deposition velocity is 1 cm/sec and the ambient concentration of NH_3 is 6 ppb. Using these figures, the deposition rate for NH_4^+ is 190 Tg(N)/yr and that for NH_3 is 575 Tg(N)/yr. If we assume that all NH_3 and NH_4^+ enters the atmosphere as NH_3, then 762 Tg(N)/yr enter the atmosphere as NH_3, of which 3.1 Tg/yr are from anthropogenic sources.

The $NO/NO_2/NO_3^-$ portion of the nitrogen cycle is constructed in a similar fashion. The total deposition rate of NO_3^-, estimated from rainfall data, is 95 Tg/yr. The deposition rate of NO_2, assuming a deposition velocity of 1 cm/sec, is 145 Tg(N)/yr. A total of 14.6 Tg(N)/yr enters the NO_2 pool from anthropogenic sources and 1.9 Tg(N)/yr enters as NO produced from N_2O. If we assume that all of the NO_3^- in the atmosphere originates from NO and NO_2, then in order to balance the total flow of

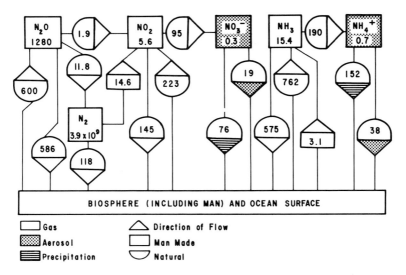

FIG. 7.3. Atmospheric nitrogen budget. The burdens are in teragrams of nitrogen and the flows are given in teragrams of nitrogen per year. NO is included in the NO_2 portion of the budget. The figures are discussed in the text.

NO_2, we must have additional natural sources amounting to 223 Tg(N)/yr. Although there are no estimates for the rate of oxidation of NO_2 to NO_3^- in the atmosphere, the auto-redox reaction $2NO_2 + H_2O \rightarrow HNO_2 + HNO_3$ may occur quite rapidly. Georgii (1963) presents evidence which suggests that subsequent oxidation of NO_2^- to NO_3^- may also proceed readily in the atmospheric environment.

It may be seen that although the estimate of the natural production of NO_x is a sensitive function of the deposition rate, which is very crude, anthropogenic NO_x is about 7% of that produced naturally. The nitrogen cycle is summarized in Fig. 7.3.

7.3 Analytical Methods

7.3.1 NITROUS OXIDE

Nitrous oxide has been analyzed using gas chromatography by a number of workers using a method developed by Bock and Schütz (1968). LaHue et al. (1970) have described some modifications of this method which make it more suitable for field work. The air stream is drawn through a series of traps the first of which removes CO_2 (using sodium asbestos or Ascarite), the second removes water (with P_2O_5 or anhydrous $CaSO_4$), and the third contains Linde molecular sieve 5A, which removes N_2O. The tube containing the molecular sieve and N_2O is connected to a gas chromatograph and the nitrous oxide is liberated by heating to 315°C. A thermal conductivity detector is used and the amount of nitrous oxide is determined by measuring the area of the detector response versus time curve. A calibration curve must of course be obtained using known nitrous oxide–air mixtures. The relative standard deviation of this method is 4.3% (Bock and Schütz, 1968).

7.3.2 NITROGEN DIOXIDE—COLORIMETRIC METHOD

The method for NO_2 determination first developed by Saltzman (1954) and later modified by Jacobs and Hochheiser (1958) has formed the basis for the accepted method for analysis of NO and NO_2. After absorption in a gas washing bottle, NO_2 forms NO_2^-, which in turn undergoes a coupling reaction with sulfanilamide and N-(1-naphthyl)-ethylenediamine to form an azo dye. Treatment by H_2O_2 reduces the interference by SO_2 and the concentration of the dye is measured colorimetrically. A calibration curve may be obtained by absorbing a known concentration of NO_2, generated by a permeation tube or a gas dilution system, or by using a standard $NaNO_2$ solution. Standardization by $NaNO_2$, although convenient, suffers from the disadvantage that nonstoichiometric amounts of NO_2^- are produced by NO_2 gas. This conversion efficiency depends on the absorbing solution and other variables. Standardization by $NaNO_2$ is recommended by the Environmental Protection Agency (1971), presumably because of difficulties encountered with NO_2 permeation tubes (Saltzman *et al.*, 1971). If it is necessary to use $NaNO_2$ for calibration of this method, the operating conditions and conversion factor used must be carefully noted.

Reagents

Absorbing reagent: 0.1 N NaOH.
Sulfanilamide reagent: One liter of a solution containing 20 g of sulfanilamide and 50 ml of concentrated H_3PO_4.
N-(1-naphthyl)-ethylenediamine dihydrogen chloride: 0.1% by weight.
Hydrogen peroxide: 0.2 ml of 30% H_2O_2 diluted to 250 ml.
Acid permanganate: Dissolve 2.5 g of $KMnO_4$ and 2.5 g of concentrated H_2SO_4 in about 90 ml of water. Dilute to 100 ml.

Dichromate oxidant: Saturate a stack of 25 sheets of 7-cm glass fiber filter paper with 25 ml of a 2.5% sodium dichromate, 2.5% sulfuric acid solution. Dry the filter at 75–80°C and store in a closed bottle. To make the NO oxidizing column, cut a filter into $\frac{1}{4}$-in. strips and place in a glass tube about 30 cm long with an inside diameter of about 1.4 cm.

Procedure

Set up a sampling train containing a pump, flowmeter, manometer, thermometer, and absorber. Add 50 ml of absorbing solution to a gas washing bottle with a fritted glass bubbler having a maximum pore diameter of 60 μm. The sampling rate should be about 0.2 liter/min. After sampling is complete, restore the volume of the absorbing solution to 50 ml. To a 10-ml aliquot of this solution add successively, with stirring, 1.0 ml of H_2O_2 solution, 10 ml of sulfanilamide solution, and 1.4 ml of the N-(1-naphthyl)-ethylenediamine solution. Allow 10 min for color development and measure the absorbance against a blank at 540 nm.

The method is sensitive only to nitrogen dioxide. Nitric oxide may be determined by either of the following methods: In the first, the sequence absorber I–$KMnO_4$–absorber II in the sampling train may be used, where absorber I collects NO_2 and absorber II collects NO_2 formed from NO. Alternatively, the dry dichromate oxidizer described by Ripley et al. (1964) may be used ahead of an absorber, which will then collect NO_2 representative of the sum $NO + NO_2$ (usually denoted NO_x). Unfortunately, the dichromate oxidizer may not be used after a bubbler in a sampling train.

7.3.3 OZONE—NEUTRAL BUFFERED POTASSIUM IODIDE

There are two principal wet chemical methods for the determination of ozone (more properly, oxidant). In both methods, iodide ion I^- is oxidized by the ozone to I_2, which is then determined colorimetrically at 352 nm. The neutral buffered potassium iodide method (Saltzman and Gilbert, 1959) has the advantage of being somewhat faster, whereas the oxidized form of iodine is more stable in the alkaline potassium iodide method. This is a distinct advantage if the sampling times are long or if it is not possible to make the spectrophotometric measurement immediately after sampling. Both methods have the disadvantage that a number of oxidizing agents, including NO_2, will also oxidize I^-, giving a positive error. Reducing agents, such as SO_2 and H_2S, will reduce ozone, giving a negative error. Substances in the distilled water which might reduce ozone should be removed by distilling the water from an all-glass distilling unit to which has been added a trace of $KMnO_4$ and $Ba(OH)_2$.

Reagents

Absorbing reagent. Make up one liter of a solution containing 13.61 g (0.1 mole) of KH_2PO_4, 14.20 g (0.1 mole) of Na_2HPO_4, and 10.0 g of KI. Age at room temperature for at least one day before use and store in an amber bottle, in a refrigerator if possible.

Standard iodine solution: Dissolve 16.0 g of KI and 3.173 g of I_2 in water, dilute to 500 ml. This solution should be stored for one day before use. Standardization of this solution may be obtained from the weight of iodine or by titration with standard sodium thiosulfate.

Procedure

Set up a sampling train with a midget impinger containing 10 ml of the absorbing solution. Sample at a 1–2 liter/min rate until adequate color has developed. The iodine color will continue to develop for about 45 min after sampling has ceased. Although it is not always convenient to wait this long (fading occurs at longer times), it is fairly important to wait the same period after sampling for all measurements. The solution volume should be adjusted to 10 ml and the absorbance measured at 352 nm against a blank which has not been exposed to the atmosphere.

Standardization is achieved using the standard iodine solution and assuming that one mole of ozone produces one mole of iodine. Although this corresponds to an overall reaction which may be written

$$2H^+ + O_3 + 2I^- \rightarrow O_2 + H_2O + I_2$$

the actual mechanism of the reaction appears to be much more involved.

Negative interferences from SO_2 and H_2S are significant because they may be very efficient and because the concentrations of these gases, particularly that of SO_2, may be the same order of magnitude as that of ozone. Saltzman and Wartburg (1965) have described a chromium trioxide column which will eliminate SO_2 without interfering with ozone.

Fifteen milliliters of solution containing 2.5 g of CrO_3 and 0.7 ml of H_2SO_4 are distributed over a 60-in.^{2} sheet of glass fiber filter paper. The sheets are folded accordianwise with $\frac{1}{4}$-in. pleats, then cut into $\frac{1}{4}$-in. strips at right angles to the pleats. The strips are placed in a U-tube, or a straight absorber about 20 mm I.D. by 15 cm long. The absorber should be conditioned so that it will not absorb ozone by passing air through it overnight.

7.3.4 OZONE—ALKALINE POTASSIUM IODIDE

The alkaline potassium iodide method for ozone is also discussed in Section 7.3.3 and in a publication of the U. S. Dept. of Health, Education, and Welfare (1965).

Absorbing reagent: Successively dissolve 40 g of NaOH and 10.0 g of KI in nearly one liter of water. Dilute to one liter.

Acidifying reagent: Dissolve 5 g of sulfamic acid in 100 ml of water, then add 84 ml of 85% phosphoric acid and dilute to 200 ml.

Standard potassium iodate: Make up one liter of a solution containing 0.758 g of KIO_3. The addition of 1.0 ml of this solution to the absorbing solution produces an amount of iodine equivalent to that produced by 400 ml of O_3 at 25°C and 1 atm.

Procedure

Use the sampling train described for the neutral buffered KI method, again using 10 ml of the absorbing solution in a midget impinger. After sampling, add water to make the volume up to 10 ml, if necessary, and store the solution in a glass-stopped graduated cylinder if the analysis is not to be completed right away. The solution should not be rinsed out of the impinger but simply allowed to drain out. To the absorbing solution, add rapidly a volume of the acidifying solution equal to one-fifth the volume of the absorbing solution, restopper the cylinder, and place it in a water bath to dissipate the heat of neutralization. As soon as the mixture is cool, transfer a portion to a cuvette, and measure the absorbance at 352 nm. The blank absorbance is determined from a sample of absorbing solution which has gone through all the steps outlined above except the exposure to air.

The calibration curve may be obtained by adding known amounts of the KIO_3 solution to the absorbing solution to make a known volume of 10 ml, followed by acidification. The IO_3^- oxidizes iodine according to

$$IO_3^- + 5I^- + 6H^+ \rightarrow 3H_2O + 3I_2.$$

It is known empirically that one mole of ozone produces 0.65 mole of I_2 under these conditions.

Interferences for this method are essentially the same as for the neutral buffered KI method. SO_2 may be eliminated by the method described previously. The use of sulfamic acid in the acidifying reagent reduces the interference by NO_2.

7.3.5 OZONE—COULOMETRIC METHOD

One of the more common instrumental methods for the determination of ozone at this date is the Coulometric method. This type of instrument usually measures the amount of current required to produce just enough hydrogen to react with the iodine formed from KI by ozone. It is also an oxidant-measuring device and has many of the same limitations as the other potassium iodide methods.

7.3.6 CHEMILUMINESCENT METHODS FOR O_3, NO, AND NO_2

A number of methods have been developed for ozone which measure the light emitted when ozone reacts with a substance. These are known as chemiluminescent methods; they rely on measurements of the light emission resulting from a chemical reaction.

One of the first of the methods in this class is that of Regener (1960, 1964), who measured the light emitted by a rhodamine-*B*-impregnated silica gel sheet. Further studies of this method by Hodgeson *et al.* (1970) have shown that it is quite specific for ozone and has a detection limit of about 10^{-3} ppm. The detector response is, however, sensitive to water vapor and the response decays with time.

There have been two other chemiluminescent methods which involve a gas-phase reaction of ozone with another molecule to produce a light-emitting species. Both of these methods appear to have the potential for measuring O_3, NO, and NO_2.

In the first of these systems, ozone in the air stream reacts with an excess of NO produced by the instrument, to form NO_2 in an excited state, which then emits light upon dropping down to a lower state (Fontijn *et al.*, 1970). The reaction system is shown in Fig. 7.4. The mixing volume is about one liter and the total pressure about 1 Torr. If NO is present in a sufficient excess, the amount of light emitted is directly proportional to the ozone concentration in the air stream. This detector may be turned around to measure NO by mixing the air stream with an excess of artificially produced ozone. In this case, the light emission will be directly proportional to the nitric oxide concentration. NO_2 may be measured by mixing the air stream with oxygen atoms which react very rapidly with NO_2 to form NO. The slower reaction of oxygen atoms with NO is also chemiluminescent and thus oxygen atoms may be used to determine NO_x. The mode using the ozone–nitric oxide reaction shows a linear response of light versus concentration over the range 4 ppb to 100 ppm.

In the second method, ozone is mixed with an excess of ethylene in a

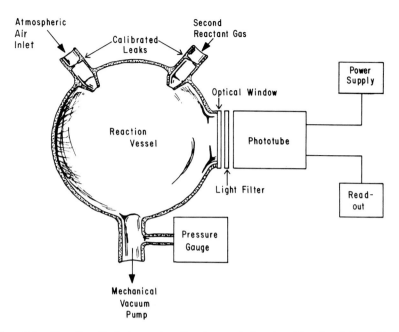

FIG. 7.4.   Reaction flask for a chemiluminescent detector. Reprinted from Fontijn *et al., Anal. Chem.* **42,** 575 (1970). Copyright by the American Chemical Society. Reprinted by permission of the copyright owner.

reaction vessel similar to that shown in Fig. 7.4, except that the total pressure is about 1 atm (Nederbragt *et al.*, 1965). While the light-emitting species has not been identified, the luminescence is again proportional to the ozone concentration. Although the original instrument was used to measure rather high ozone concentrations and had a detection limit of about 0.05 ppm, the design is being refined and one might expect an instrument to become available with a lower detection limit. Kummer *et al.* (1971) have shown that the use of gases other than ethylene can increase the sensitivity of this method. This instrument may also be used to measure NO by mixing a known, constant amount of ozone with the air stream to quantitatively react with the NO. The reduction in luminescence due to the ozone–ethylene reaction is then a measure of the amount of NO present.

PROBLEMS

1. In the colorimetric determination of NO_2, the ratio of absorbance to concentration of the azo dye is about 9×10^4 (m/liter)$^{-1}$ when an

absorption cell 1 cm long is used. Considering the method outlined in Section 7.3.2, how long would air have to be drawn through the absorbing solution at a rate of 0.2 liters/min to yield a transmittance of 44% for an NO_2 concentration of 0.1 ppm (25°C, 1 atm)? Assume a conversion efficiency for NO_2 to NO_2^- of 35%.   [53 hr]

2. The determination of NO_2 outlined in Problem 1 is to be calibrated by standard $NaNO_2$ solutions. In this procedure, 1 ml of standard $NaNO_2$ solution will be diluted to 10 ml with the absorbing solution. The azo dye will be formed by adding the other reagents in the usual order. Calculate the range of concentrations of the $NaNO_2$ solution required for this calibration. (The range of concentrations of $NaNO_2$ which will give transmittances in the range 99% to 1%.)   [1.1 × 10^{-6} to 4.9 × 10^{-4} m/liter]

3. Bodenstein (1918) proposed the following mechanism to account for the thermal oxidation of nitric oxide ($2NO + O_2 \rightarrow 2NO_2$):

(1) $NO + O_2 \rightarrow NO_3$ k_1

(2) $NO_3 \rightarrow NO + O_2$ k_2

(3) $NO_3 + NO \rightarrow 2NO_2$ k_3

This reaction follows the rate expression $d(NO_2)/dt = 2k_r(NO)^2(O_2)$ over a large range of pressure. What approximations must be made to obtain this rate expression? How is k_r related to k_1, k_2, and k_3?   [$k_r = k_1k_3/k_2$, if $k_2 \gg k_3(NO)$]

4. Glasson and Tuesday (1963) obtained the value 4.7 × 10^{-8} ppm^{-2} hr^{-1} for k_r in Problem 3 at about 25°C. Schott and Davidson (1958) found that the rate constant k_3 was given by the expression $6 \times 10^{13}e^{-1400/RT}$ cm^3 mole^{-1} sec^{-1}. Calculate the concentration of NO_3 in an atmosphere containing 0.05 ppm NO and 0.2 atm O_2.   [$(NO_3) = 5.4 \times 10^{-10}$ ppm $k_3 = 8.4 \times 10^5$ ppm^{-1} hr^{-1}]

5. Combine the answers given above with the equilibrium constant expression given by Schott and Davidson for the reaction

$N_2O_5 \leftrightarrow NO_2 + NO_3$, $K_{eq} = 10^{4.97}e^{-20,100/RT}$ moles/liter

and calculate the concentration of N_2O_5 in equilibrium with 0.05 ppm each of NO and NO_2.   [$(N_2O_5) = 6.2 \times 10^{-9}$ ppm]

6. Calculate the mass of the atmosphere and the number of moles of gas in the atmosphere from the following data.* Assume that the surface of the earth is 5.10 × 10^8 km^2 and that the density of the atmosphere

* U.S. Comm. on Extension to the Standard Atmosphere (1962).

is independent of latitude at constant altitude. What fraction of the mass is in the troposphere (below 11 km)?

Altitude (km)	Density (kg/m³)	Altitude (km)	Density (kg/m³)
0	1.225	20	8.89×10^{-2}
2	1.007	25	4.01×10^{-2}
4	0.819	30	1.84×10^{-2}
6	0.660	40	4.00×10^{-3}
8	0.526	50	1.03×10^{-3}
10	0.414	60	3.1×10^{-4}
12	0.312	70	8.8×10^{-5}
14	0.228	80	2.0×10^{-5}
16	0.1665	90	3.2×10^{-6}
18	0.1217		

REFERENCES

Altshuller, A. P. (1958). *Tellus* **10**, 479.
Altshuller, A. P., and Bufalini, J. J. (1965). *Photochem. and Photobiol.* **4**, 97.
Altshuller, A. P., and Bufalini, J. J. (1971). *Environ. Sci. Technol.* **5**, 39.
Arnold, P. W. (1954). *J. Soil Sci.* **5**, 116.
Bates, D. R., and Hays, D. B. (1967). *Planet. Space Sci.* **15**, 189.
Bock, R., and Schütz, K. (1968). *Fresenius' Z. Anal. Chem.* **237**, 321.
Bodenstein, M. (1918). *Z. Elektrochem.* **24**, 183.
Clyne, M. A. A., Thrush, B. A., and Wayne, R. P. (1964). *Trans. Faraday Soc.* **60**, 359.
Coomber, J. W., and Pitts, J. N. (1970). *Environ. Sci. Technol.* **4**, 506.
Craig, R. A. (1965). "The Upper Atmosphere." Academic Press, New York.
Delwiche, C. C. (1970). *In* "The Biosphere," Scientific American Book. Freeman, San Francisco, California.
Environmental Protection Agency (1971). *Fed. Regist.* **36**, 8186.
Fontijn, A., Sabadell, A. J., and Ronco, R. J. (1970). *Anal. Chem.* **42**, 575.
Friedlander, S. K., and Seinfeld, J. H. (1969). *Environ. Sci. Technol.* **3**, 1175.
Georgii, H. W. (1963). *J. Geophys. Res.* **68**, 3963.
Glasson, W. A., and Tuesday, C. S. (1963). *J. Amer. Chem. Soc.* **85**, 2901.
Harrison, H. (1970). *Science* **170**, 734.
Healy, T. V., McKay, H. A. C., Pilbeam, A., and Scargill, D. (1970). *J. Geophys. Res.* **75**, 2317.
Hodgeson, J. A., Krost, K. J., O'Keeffe, A. E., and Stevens, R. K. (1970). *Anal. Chem.* **42**, 1795.
Jacobs, M. B., and Hochheiser, S. (1958). *Anal. Chem.* **30**, 426.
Johnston, H. (1971). *Science* **173**, 517.
Kummer, W. A., Pitts, J. N., and Steer, R. P. (1971). *Environ. Sci. Technol.* **5**, 1045.
LaHue, M. D., Pate, J. B., and Lodge, J. P. (1970). *J. Geophys. Res.* **75**, 2922.
Leighton, P. A. (1961). "Photochemistry of Air Pollution." Academic Press, New York.

Lowry, T., and Schuman, L. M. (1956). *J. Amer. Med. Ass.* **162,** 153.

Nederbragt, G. W., van der Horst, A., and van Diujn, J. (1965). *Nature (London)* **206,** 87.

Regener, V. H. (1957). *J. Geophys. Res.* **62,** 221.

Regener, V. H. (1960). *J. Geophys. Res.* **65,** 3975.

Regener, V. H. (1964). *J. Geophys. Res.* **69,** 3795.

Ripley, D. L., Clingenpeel, J. M., and Hurn, R. W. (1964). *Intern. J. Air Pollut.* **8,** 455.

Robinson, E., and Robbins, R. C. (1970). *J. Air Pollut. Contr. Ass.* **20,** 303.

Saltzman, B. E. (1954). *Anal. Chem.* **26,** 1949.

Saltzman, B. E., and Gilbert, N. (1959). *Anal. Chem.* **31,** 1914.

Saltzman, B. E., and Wartburg, A. F. (1965). *Anal. Chem.* **37,** 779.

Saltzman, B. E., Burg, W. R., and Ramaswamy, G. (1971). *Environ. Sci. Technol.* **5,** 1121.

Schott, G., and Davidson, N. (1958). *J. Amer. Chem. Soc.* **80,** 1841.

Schuck, E. A. and Stephens, E. R. (1969). *In* "Advances in Environmental Sciences and Technology," Volume I (J. N. Pitts and R. L. Metcalf, eds.). Wiley (Interscience), New York.

Schütz, K., Junge, C., Beck, R., and Albrecht, B. (1970). *J. Geophys. Res.* **75,** 2230.

Stephens, E. R. (1967). *Atmos. Environ.* **1,** 19.

U.S. Comm. on Extension to the Standard Atmosphere (1962). "U.S. Standard Atmosphere, 1962." U.S. Govt. Printing Office, Washington, D.C.

U.S. Dept. of Health, Education, and Welfare (1965). Selected methods for the measurement of air pollutants. Publ. 999-AP-11. Publ. Health Serv., Cincinnati, Ohio.

Westberg, K., Cohen, N., and Wilson, K. W. (1971). *Science* **171,** 1013.

CHAPTER 8

CARBON COMPOUNDS

There are two oxides of carbon of significance in the chemistry of air. Carbon dioxide (CO_2) is the most concentrated natural trace substance in air after argon and water vapor. Besides its natural existence, human activity in the form of fossil fuel combustion (petroleum gas and coal) is altering the atmosphere and hydrosphere. Carbon monoxide (CO) is also found naturally, but in much lower concentrations. In urban areas, CO is often the most concentrated trace constituent after CO_2. Its toxicity to warmblooded animals makes it a pollutant of importance as well. This chapter will summarize the main features of the chemistry of carbon oxides in air, including sources, atmospheric behavior, sinks, and measurement.

Besides the carbon oxides, there is an almost endless list of other carbon containing compounds which play important roles in the atmosphere. Although their presence is easily recognized, the complexity of analysis, the infinite possibilities for molecular variations, and the lack of communication between organic chemists and the more physically oriented atmospheric scientists has led to a lack of a well-developed body of information on carbon compounds in the air. To be sure, there have been studies of such molecules as methane, formaldehyde, peroxyacyl nitrates, polynuclear aromatics, terpenes, and other materials suspected of being important in one or another form of air pollution. However, in view of the breadth of organic chemistry, the field of study of these compounds in the atmosphere is essentially a virgin one.

Our purpose here will be to focus on the well-developed, basic aspects of carbon compounds in air, with only a mention of the less-established, current research topics.

8.1 Sources and Sinks of Carbon Dioxide

Carbon dioxide is injected into the atmosphere both in a natural cycle and by human activity. Most of the CO_2 that is produced annually is natural, with man's contribution currently representing about 4% of the total production. In order to put this human production in perspective, it is useful to take a geochemical approach. Figure 8.1 shows the terrestrial carbon cycle without quantities attached to the flows of material. Figure 8.3, which is given later, provides quantitative estimates of these.

Carbon dioxide is produced naturally by respiration and decay of plants and humus, a cycle that is almost perfectly balanced. Man produces much smaller quantities by combustion of carbonaceous materials. In the early states of the formation of the earth, the same oxidation occurred, yielding CO_2 as a natural constituent of the atmosphere. Over the eons since the formation of the earth, most of the primordeal CO_2 has been trapped in the lithosphere as limestone ($CaCO_3$) and other carbonates. Some CO_2 is dissolved in the oceans as carbonic acid and its dissociation products. Since CO_2 solubility in H_2O is both pH- and temperature-dependent, and since carbonate rocks do dissolve in some circumstances (for instance, in the weathering of mountains), there is a secondary source of CO_2 of potentially large magnitude.

Figure 8.1 depicts the carbon cycle with special emphasis on those cycles that affect the atmosphere. The amounts of carbon in the earth "reservoir" are shown relative to atmospheric CO_2 carbon to illustrate how little of the carbon in the cycle is in the atmosphere.

It is difficult to model this system mathematically, due to the uncertainty of the rates of transport of carbon via the many processes that are listed. Hence, it is difficult to predict the increase of the atmospheric CO_2 average concentration due to increase in one or another source of CO_2. There are uncertainties in the effects of small changes in the large amounts of carbon in reservoirs other than the atmosphere. For instance, a small increase in the average temperature of the ocean deeps would result in a decreased solubility of CO_2 and ultimately an increase in atmospheric CO_2.

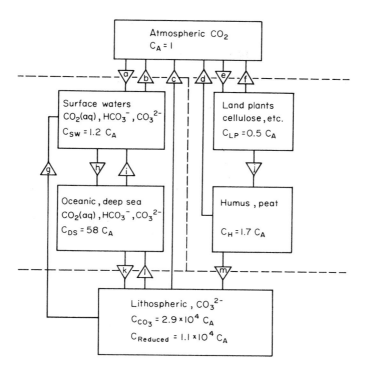

FIG. 8.1.   The carbon cycle. The amounts of carbon in each reservoir (rectangle) are shown relative to that in the atmosphere (C_A): The numerical estimates are from NAS–NRC (1966). The processes are shown as triangles: (a) air–sea sorption of CO_2; (b) water–air desorption of CO_2; (c) combustion; (d) oxidation of dead plants; (e) photosynthetic CO_2 consumption; (f) respiration; (g) rock weathering; (h) ocean surface–deep sea exchange; (i) deep sea–ocean surface exchange; (j) decay of plant material; (k) oceanic bottom deposition; (l) dissolution of marine deposits; (m) metamorphosis.

The magnitude of such an effect remains unknown due to the complexity of the overall system.

There is, however, another approach to understanding the dynamics of such complex systems. If known changes occur in part of a system, the response can be observed in the system as a whole, providing clues about the unknown factors that precluded an *a priori* approach. The terrestrial CO_2 system is perhaps a unique example of this sort of phenomenon, both with regard to complexity and its physically large size.

The specific perturbation that has occurred is anthropogenic combustion of fossil fuels. (Fossil fuels are all such things as coal, oil, natural gas, etc.,— fossilized plants with a high carbon content.) The change in global CO_2

production rate—process c in Fig. 8.1—is sufficiently large to result in shifts in several other *flows* in the system, and hence also changes in the *amounts* of carbon in several of the reservoirs.

Keeling and Pales (1965), SMIC (1971), and Keeling (1971) have shown that since 1958, atmospheric CO_2 has increased at a rate of a little less than 1 ppm per year on a global basis. Since the present CO_2 content is about 325 ppm, this change is of the order of 0.3% per year. Figure 8.2 shows the CO_2 data taken at the Mauna Loa Observatory (MLO) at an altitude of 3400 m above sea level which shows this striking measured increase in annual average CO_2 in the decade 1960–1970. Earlier CO_2 analyses are somewhat questionable because of a lack of standards and the necessity for wet chemical methods. However, Callendar (1958) shows a large amount of data that yield an average for the period before 1901 of about 290 ppm. This increase of ~35 ppm has been attributed to human activity in spite of other possibilities, which will be discussed. The increased CO_2 could be due to the following factors.

1.   A decrease in the CO_2 in surface waters, or a decrease in the rate of sorption by the ocean surface (e.g., due to temperature increase).

2.   An increase in oxidation of dead plants.

3.   A decrease in photosynthesis rate or an increase in respiration by plants.

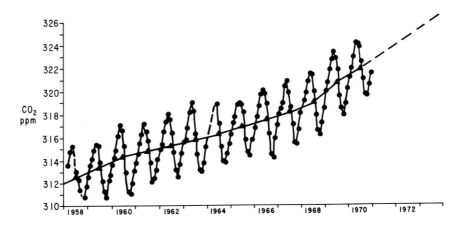

Fig. 8.2.   Carbon dioxide concentration (ppm) as a function of time for the period 1958–1970 at Mauna Loa Observatory, Hawaii (3400 m altitude). (●) Monthly averages; (▲) annual averages of monthly averages. Data in period 1958–1963 are from Pales and Keeling (1965). Those in period 1964–1970 are from Keeling (1971).

4.   An increase in the oxidation of lithospheric, reduced carbon due to human use of fossil fuels.

There are several reasons why the last item appears to dominate the others.

1.   No extensive changes have been noted in the biosphere to suggest that there is now more dead plant material to oxidize or that there is a systematic global decrease in the rate of photosynthesis.

2.   The temperature of the oceans has not changed enough to account for the observed CO_2 effects. The Northern Hemisphere average temperature has increased less than 0.5°C since 1900, a figure that should relate closely to the temperature of the mixed layer of the oceans. This temperature change, combined with the temperature dependence of Henry's law, predicts less than 2.5% increase in CO_2 (Tukey, 1965).

However, the solubility of CO_2 in seawater is a complicated function of pH, (HCO_3^-) ion concentration, and biological activity, so that it is hazardous to carry this approach further without more information.

3.   The clincher is that because fossil fuels have essentially no ^{14}C, the $^{12}CO_2/^{14}CO_2$ ratio in the atmosphere changes as fossil carbon is burned.* This dilution of $^{14}CO_2$ is known as the "Suess effect," after Professor Hans Suess, who first observed it in studies of isotope ratios in tree rings (Revelle and Suess, 1957). The result of these considerations is a partitioning of the fossil CO_2 between those reservoirs that respond to such rapid ($t < 100$ year) changes, namely the atmosphere, the biosphere, and the surface waters of the oceans. An estimate can then be made of the amount of the increase in CO_2 that is due to combustion of fossil fuels. The quantity so arrived at is comparable to that predicted from records of fuel consumed.

Assuming on this basis that the major change in the carbon cycle since the industrial revolution of the 19th century is the introduction of fossil fuel combustion, it is possible to construct a more quantitative version of the atmospheric part of the cycle. Figure 8.3 is such a cycle, with flow strengths and burdens included. Consideration of Figs. 8.1 and 8.2 suggests that slightly less than half of the 5000 Tg/yr of carbon injected by man into the atmosphere (as CO_2) remains there and half is removed to other reservoirs. The 5000 Tg/yr figure is based on the figures for world production of fossil fuels (Bolin, 1970). The net increase of ca 2000 Tg/yr of carbon (as CO_2) in the atmosphere amounts to about 0.3%/yr or about 1 ppm/yr.

---

* $^{14}CO_2$ is produced naturally by a nuclear reaction between cosmic ray neutrons and nitrogen in the upper atmosphere and is also left over after nuclear explosions.

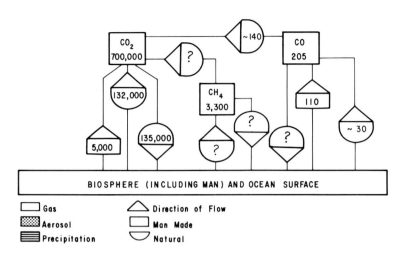

F<small>IG</small>. 8.3.   The atmospheric portion of the carbon cycle. The burdens are given in teragrams of carbon; the flows are expressed in teragrams of carbon per year.

Another aspect of Fig. 8.2 bears mention, namely the sinusoidal annual variation of CO_2. One may ascribe the decrease (in the Northern Hemisphere summer) to increased photosynthesis and the increase to increased use of fuels for heating (in Northern Hemisphere winter). A lack of any effect due to the Southern Hemisphere can be assumed to be due to slow mixing across latitude lines as well as low population and smaller amounts of land biosphere in the Southern compared to the Northern Hemisphere. Details of the phase relationship and altitude dependence have been studied by Bolin and Bischof (1970).

So far, we have discussed only the global aspects of CO_2 in the atmosphere, and its changes on a year or greater time scale. If a smaller spatial scale is chosen—say the distance across an urban area—then the CO_2 time dependence changes dramatically. Most large sources of CO_2 (power plants) are located in or near urban areas, so that perhaps 1% or less of the area of the earth is effected more directly by the sources. The average CO_2 content in urban areas should thus be considerably above the global average. Also, due to the proximity of the sources, the time variability should be much greater. Unfortunately, few records of urban CO_2 have been taken regularly, so that it is not possible to provide a representative picture. Concentrations of 500–1000 ppm are frequently found in New York City for periods of a fraction of a day. It is interesting to consider the fact that several large areas of the earth are heavily urbanized (e.g., the East Coast of the United States, Central Europe, and Japan), and that environmental effects due to an increasing CO_2 concentration might be expected to occur on a meso scale or synoptic scale before they happen globally.

8.2 Sources and Sinks of Carbon Monoxide

There are two basic classes of CO production, natural and anthropogenic. The natural production includes both ^{12}CO and very small amount of ^{14}CO, while that produced by man is dominated by fossil fuel combustion and hence by ^{12}CO. Junge et al. (1971) suggest that man produces about 70% of the total atmospheric CO each year or 2.6×10^{14} g of CO per year. As carbon, this amounts to 110 Tg/yr, as shown in Fig. 8.3. The natural production rate is somewhat less certain than the anthropogenic rate, but is about 30 Tg of carbon per year (Junge et al., 1971). It must be emphasized that these figures are approximate and speculative. The natural oceanic source is apparently biological in nature.

There are at least two known CO sinks: (a) reaction to produce CO_2 in the lower stratosphere, and (2) biological removal in soil. There may be other sinks, as yet unidentified. The strength of these sinks can only be deduced from many assumptions, and those values given in Fig. 8.3 are intended mainly for illustration of a balanced budget, which is not necessarily the case. The 140 Tg/yr figure is derived from the estimate by Junge et al. (1971) of the value for the marine troposphere flux.

It is interesting to note that the anthropogenic production of CO is almost totally within small land areas of the Northern Hemisphere. Thus, while natural sources may be very important on a global scale, they are probably quite insignificant in a polluted city such as New York or Los Angeles. Indeed, it is very safe to assume that the CO concentration in most urban locations is essentially 100% anthropogenic. This is evidenced by typical "background" levels of CO around 0.1–0.2 ppm, with urban levels frequently in the 10–20 ppm range.

The main anthropogenic CO source is the internal combustion engine of automobiles and trucks which accounts for $\sim$58% of the production (U. S. Dept. of Health, Education, and Welfare, 1970a). The rest of the sources are divided among stationary sources (2%), industrial processes (11%), garbage incineration (8%), and miscellaneous sources, including forest firest, aircraft, etc. (21%).

In all combustion sources of CO, the ratio of carbonaceous fuel to oxygen is either too high to permit the complete formation of CO_2, or the temperature is too low to permit the oxidation to occur. The automotive source is characterized by a combustion chamber with cool walls where the oxidation process is slowed down, thus promoting CO formation.

The concentration of CO in cities is highly variable and depends on time and location with respect to sources (streets). Usually, it is possible to observe a diurnal pattern that is highly correlated to the local traffic. If measurements are made immediately downwind of a street, the concentration of CO is dominated by the existing traffic and it is often possible to even observe plumes from individual vehicles.

Figure 8.4 shows typical CO levels in Los Angeles, Detroit, and Manhattan, as a function of time of day. These data consist of the average of a large number of hourly averages for a given time of day plotted versus time. No one day would necessarily follow this curve, although there is often a recognizable pattern in the record of a single day.

The levels reached in these records lie in the range of $1 \leq [CO] \leq 20$ ppm, which deliniates the typical situation in urban areas today.

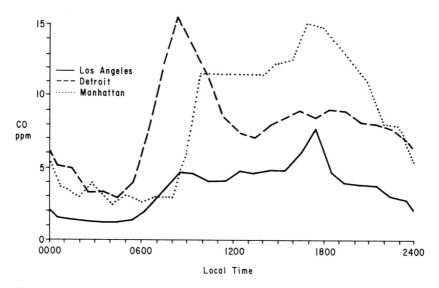

FIG. 8.4.   Carbon monoxide as a function of time of day in three cities. Values are averages for weekdays. Data for Manhattan from Johnson *et al.*, *Science* **160**, 67 (1968); data for Los Angeles and Detroit from Colucci and Begeman (1969); used with permission.

8.3 Behavior of Carbon Oxides in the Atmosphere

Both CO and CO_2 are—at atmospheric levels—tasteless, odorless, and colorless gases. Both are relatively unreactive in comparison to either sulfur or nitrogen oxides. The residence times in the atmosphere are estimated to be ~ 0.1 yr for CO (Junge *et al.*, 1971) and perhaps 2 yr for CO_2 (SMIC, 1971). Thus it is safe to assume in many cases that both CO and CO_2 are conserved in distances of the scale of a city or times on the order of a few hours. The possibility exists that CO may be involved in photochemical smog, however, making the actual lifetime in such situations somewhat uncertain.

There is another gross similarity between CO and CO_2 in that both seem to be highly involved (on a global scale) with biological activity. However, the similarities cease at this point.

Carbon dioxide is present in a sufficient quantity to play a role in the heat balance of earth, both by absorption and emission of infrared radiation (Fleagle and Businger, 1963). Due to its smaller concentration, CO does not play such a role. Carbon dioxide also is involved in the carbonate balance of natural waters, and is responsible for a small degree of acidification of rainwater.

Of all global pollution problems, the CO_2 increase is certainly the best monitored and probably the best understood in terms of its effects. The consensus of climatologists is that if the CO_2 increase continues, the first effect will be a very small increase in the surface temperature of the earth. No increase in planetary temperature is forecast since no albedo change is assumed. SMIC (1971) suggests an $\sim 0.5°C$ increase in average surface temperature by the year 2000, all other factors being held constant (such as cloudiness, albedo, and other climatic variables). A doubling of CO_2 to ~ 650 ppm might increase the temperature by 2°C.

At the same time, increased CO_2 can be expected to directly affect the biosphere by increasing the rate of photosynthesis in both land and ocean plants. The increased temperature may be expected to have effects also, but these are far more difficult to predict.

While CO_2 is not considered toxic in levels forecast for the end of the century, or even in much higher concentrations, CO has decidedly deleterious effects on warm-blooded animals. The basis of the effect is the reaction of CO with hemoglobin to form carboxy hemoglobin. The result of this reaction is a reduced capacity of blood to carry oxygen, so that acute CO poisoning results in asphyxiation. There are, however, effects at lower levels. Long-term exposures to 100 ppm can result in behavioral changes, decreased visual performance, cardiovascular effects, and possibly other deleterious responses. Lower level exposures may have an important impact, but proof via epidemiological statistics is difficult (U. S. Dept. of Health, Education, and Welfare, 1970a; Hexter and Goldsmith, 1971).

We stated earlier that in most cases CO could be considered as unreactive in air, and this is probably true for the bulk of the troposphere. However, there are two situations where CO plays an important role in photochemistry: in the stratosphere and in photochemical smog. The degree to which CO is involved is still a topic of extensive research, but its function in the reaction systems seems established (Pressman and Warneck, 1970; Westberg et al., 1971).

In both situations, the CO is oxidized to CO_2 by the free radical OH, as mentioned earlier in Chapter 7:

$$CO + OH \rightarrow CO_2 + H$$

This reaction is probably important in the CO sink mechanism in the stratosphere, and may be important as an *in situ* sink in urban smog. Just as it is not possible to provide a quantitative picture for the overall photochemical smog mechanism, the exact importance of this reaction is not known.

8.4 Other Carbon Compounds

As mentioned earlier, the study of organic compounds in air is poorly developed. No well-defined cycles can be proposed for most cases, and quantitative aspects are almost absent. Nonetheless, there are a wide variety of organic molecules in air. We will only be able to list some of the more notable ones along with their sources, starting with naturally occurring substances and progressing to anthropogenic materials.

Most natural organic materials seem to arise from biological sources. A few come from naturally occurring fires, such as lightning-ignited forest fires. The aroma of a forest, the scent of a flower, the stench of rotting vegetation, and the odors of the seashore are all involved with natural organic materials emitted from the biosphere.

8.4.1 METHANE

The most abundant single organic material is the simplest hydrocarbon, methane, CH_4. Natural levels of CH_4 in the troposphere range between 1 and 2 ppm. Since it is a gas of rather low reactivity, its lifetime in air is probably substantial and could be comparable to or larger than that of CO. Methane is ubiquitous in the troposphere as a result of this low reactivity. The sources of CH_4 are possibly dominated by microbiological processes such as the rotting of dead plants in swamps, bogs, and paddy fields. Natural leaks from gas pockets associated with fossil fuel deposits might be important. Koyama (1963) estimates the global emission at 300 Tg/yr, which Robinson and Robbins (1968) suggest is too low. Swinnerton et al. (1969) found atmospheric methane to be equilibrated with that in Atlantic seawater. Some methane is produced by human activity as well.

The sinks for CH_4 are not well known, but probably include biological activity at the earth's surface as well as reaction in the stratosphere. No quantities are shown in Fig. 8.3, due to this lack of data.

8.4.2 TERPENES

Many plant species emit terpenoid compounds as a natural part of their life processes. Additional terpenes are emitted when the plant is mechanically damaged or upon defoliation of deciduous trees. The rates of emission are difficult to estimate, due to the variations among species, the dependence on climatic factors and sunlight, the season, etc. Rasmussen (1969) identifies α-pinene, β-pinene, myrcene, limonene, and isoprene as the dominant compounds emitted in forests in the United States. These compounds all contain one or more carbon–carbon double bonds and are very reactive in the presence of ozone. α-Pinene is one of the most abundant and has exhibited a tendency to form aerosols in air by photochemical processes.

The role of terpenes and other plant emissions in air chemistry and especially in the production of aerosols is not known in detail. The possibility exists that such compounds are a leading source of hazes in sparcely populated continental areas (Went, 1966). Ripperton et al. (1967) suggest that terpenes are a major sink for the ozone carried to the troposphere from the stratosphere.

α-pinene   isoprene

It is interesting to compare the annual estimates of global emissions and of atmospheric concentrations for the various terpenes with those for methane (U. S. Dept. of Health, Education, and Welfare, 1970b). Rasmussen and Went (1965) suggest terpene emissions that are comparable on a mass basis to those of methane. Yet the atmospheric levels of terpenes ($\sim$50 $\mu g/m^3$) are much lower than that of methane ($\sim$700 $\mu g/m^3$). This suggests that the lifetime of the terpenes is much shorter than that of methane. In view of the reactivity of olefins with O_3 and other known atmospheric constituents, this seems reasonable.

8.4.3 OTHER NATURAL ORGANIC COMPOUNDS

Besides methane and terpenes, there are many other organic compounds that are important in the chemistry of air. Most of these have received little attention and not much is known about their sources and sinks, role in atmospheric processes, lifetime, and atmospheric concentrations. The list in Table 8.1 intended to call attention to the existence of these substances.

8.4.4 ANTHROPOGENIC HYDROCARBONS

While the global natural emission of nonmethane hydrocarbons may be of the order of 400 Tg/yr, the emission from all human activity in the U. S.

TABLE 8.1

Compound	Source	Obvious feature or function
Sulfur-containing:		
Methyl mercaptan	Rotting biota, seashore	Odor
Dimethyl sulfide	Rotting biota, seashore	Odor
Halogen-containing:		
Methyl iodide	Seaweed	Old commercial source of iodine
Metal–organic compounds:		
	?	?
Methyl mercury	?	?
Particulate matter:		
Microorganisms (bacteria yeasts, molds, fungi, etc.)	?	Atmospheric transport of pathogens
Plant pollen	Trees, weeds, flowers, etc.	Pollination
Soot and partially burned plant material	Forest fires	Visibility degradation, long-distance transport

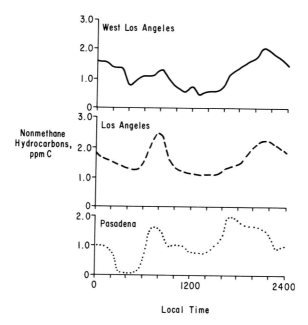

FIG. 8.5. Total hydrocarbons sensed by flame ionization for three sites in Los Angeles County. Five months of hourly averages are included. An assumed background level of methane was subtracted.

is estimated to be about 32 Tg/yr (U. S. Dept. of Health, Education, and Welfare, 1970b). Most (49%) of this emission is from motor vehicles, ∼2% is from stationary fuel combustion, 10% is from solvent evaporation, and the remainder is from industrial processes (14%), solid waste disposal (5%), gasoline marketing (4%), forest fires (7%), and other miscellaneous sources. It is obvious from these figures that urban hydrocarbon levels are dominated totally by the local sources, especially automobiles.

These emissions lead to ambient levels in the range from background up to a few ppm. Most measurements of hydrocarbons in cities are made with a nonspecific method (flame ionization, to be discussed later), so that detailed composition data are rare. However, as expected from the vehicular source, total hydrocarbons in cities often exhibit a diurnal pattern similar to CO. Often, a set amount (say 1 ppm) is subtracted from the measured quantity to compensate for natural CH_4. Reported data are then assumed to be nonnatural methane and other hydrocarbons, and are reported as ppm C. Figure 8.5 shows typical averages of hourly averages of

TABLE 8.2

Compound	Concentration	
	ppm	ppm (as carbon)
Methane	3.22	3.22
Ethane	0.098	0.20
Propane	0.049	0.15
Isobutane	0.013	0.05
n-Butane	0.064	0.26
Isopentane	0.043	0.21
n-Pentane	0.035	0.18
2,2-Dimethylbutane	0.0012	0.01
2,3-Dimethylbutane	0.014	0.08
Cyclopentane	0.004	0.02
3-Methylpentane	0.008	0.05
n-Hexane	0.012	0.07
Total alkanes (excluding methane)	0.3412	1.28
Ethylene	0.060	0.12
Propene	0.018	0.05
1-Butene + Isobutylene	0.007	0.03
trans-2-Butene	0.0014	0.01
cis-2-Butene	0.0012	Negligible
1-Pentene	0.002	0.01
2-Methyl-1-butene	0.002	0.01
trans-2-pentene	0.003	0.02
cis-2-pentene	0.0013	0.01
2-Methyl-2-butene	0.004	0.02
Propadiene	0.0001	Negligible
1,3-Butadiene	0.002	0.01
Total alkenes	0.1020	0.29
Acetylene	0.039	0.08
Methylacetylene	0.0014	Negligible
Total acetylene	0.0404	0.08
Benzene	0.032	0.19
Toluene	0.053	0.37
Total aromatics	0.0850	0.56
Total, all hydrocarbons	3.7886	5.43

TABLE 8.3

AVERAGE AND HIGHEST CONCENTRATION MEASURED FOR
VARIOUS AROMATIC HYDROCARBONS IN LOS ANGELES, 26 DAYS,
SEPTEMBER–NOVEMBER, 1966[a]

Aromatic hydrocarbon	Average concentration (ppm)	Highest measured concentration (ppm)
Benzene	0.015	0.057
Toluene	0.037	0.129
Ethylbenzene	0.006	0.022
p-Xylene	0.006	0.025
m-Xylene	0.016	0.061
o-Xylene	0.008	0.033
i-Propylbenzene	0.003	0.012
n-Propylbenzene	0.002	0.006
3- and 4-Ethyltoluene	0.008	0.027
1,3,5-Trimethylbenzene	0.003	0.011
1,2,4-Trimethylbenzene, and i-butyl- and sec-butylbenzene	0.009	0.030
tert-Butylbenzene	0.002	0.006
Total aromatics	0.106	0.330

[a] Adapted from Lonneman et al. (1968).

data taken in Los Angeles County to illustrate the level and diurnal pattern exhibited in "nonmethane" hydrocarbons.

As might be suspected from the nature of the sources, the variety of hydrocarbons present in urban air is large, but is characterized by a dominance of petroleum distillates. Average compositions of Los Angeles hydrocarbon samples are summarized in Table 8.2. Maximum concentrations for aromatic compounds are also summarized for this locale in Table 8.3.

The concentrations of gaseous hydrocarbons thus defined clearly outweigh the organic matter found in aerosols. This result is similar to that for other gaseous pollutants such as SO_2 and NO_2.

8.4.5 SECONDARY PRODUCTS IN PHOTOCHEMICAL SMOG

The atmosphere in places that are frequently effected by photochemical smog develops substantial concentrations of organic materials that are created in the atmosphere from the primary pollutants injected by the sources. The list of these compounds produced in the "atmospheric reaction vessel" is lengthy, but there are a few noteworthy substances.

Carbonyl compounds—mainly aldehydes—are frequently found to exist in concentrations around 0.1 ppm. The peak concentration of these substances seems to coincide with the ozone maximum, suggesting a photochemical source. One reaction which could lead to the production of aldehydes is the ozonolysis of olefinic hydrocarbons. Some analyses determine total aldehydes, and at other times, specific compounds are detected or measured. Two substances for which analytical procedures have been developed and data compiled are formaldehyde ($H_2C{=}O$) and acrolein

$$\left(CH_2{=}CH{-}C \begin{smallmatrix} \displaystyle\diagup\!\!\diagup O \\ \\ \diagdown H \end{smallmatrix} \right)$$

Both are found frequently in photochemical smog. There exist observations of total aldehydes that reached an hourly average of 1.3 ppm, indicating that carbonyl compounds may be one of the main byproducts of the reaction of hydrocarbons.

Aerosols are formed from some hydrocarbons in the presence of nitrogen oxides, ozone, sulfur dioxide, and sunlight. The mechanism (more properly, the group of mechanisms) by which particulate matter is formed or grows on existing nuclei is not known. Table 8.4 shows a typical analysis of material collected on filters in Los Angeles smog, including noncarbonate carbon. Since no molecular information was obtained, we can only speculate about the nature of the organic particulate matter; however, it is clear that the organic fraction was the largest single component of the samples. Three facts emphasize this dominance.

First, the carbon is bound to hydrogen, perhaps oxygen, nitrogen, sulfur, and other atoms. Thus the 26% figure is very conservative since these additional atoms would raise the percentage. (The analyses usually account for around 60% of the total weighable material.)

Second, we know that many organic materials in particulate form are either volatile or liquid (or both) and will not stay on a filter after sampling. They may volatilize when exposed to the reduced pressures behind the front surface of the filter or they may just physically flow through the filter. As much as 50–75% of the material present in air might volatilize upon removal by filters or impaction samplers.

Third, there are not many large sources of carbonaceous particulate matter known to exist in places like Los Angeles (Miller *et al.*, 1972).

TABLE 8.4

PERCENTAGE BY WEIGHT OF SEVERAL ELEMENTS IN AEROSOL
COLLECTED IN PASADENA, CALIFORNIA, 22 AUGUST 1971,
1200–1600 PDT[a]

Element	Percentage	Element	Percentage
Al	1.73	Mg	2.3
Ba	0.02	Mn	0.04
Br	0.8	Na	1.11
C (non-carbonate)	26.0[b]	NH_4^+	—
Ca	3.8	NO_3^-	—
Cl	0.39	Pb	5.4
Cu	0.18	Si (insoluble)	—
Fe	8.0	Si (soluble)	—
I	0.011	SO_4^{2-}	—[c]
In	0.0002	V	0.01
K	0.51	Zn	0.73

[a] Mueller (1971).
[b] Carbonate is usually low.
[c] Typically 5–10%.

Thus, the importance of particulate matter formed in air out of hydrocarbons seems well established. This matter is currently the subject of intensive research which, due to the complex nature of the problem, can be expected to continue for some time.

Peroxy acyl nitrates (PAN) were mentioned in Chapter 7 and are another example of a secondary organic material generated in the atmosphere (U. S. Dept. of Health, Education, and Welfare, 1970c).

8.4.6   ORGANIC SULFUR COMPOUNDS

Although they usually are negligible in concentration compared to SO_2 or H_2S, organic sulfur compounds from some human activities are very important because of their strong odor. Many of the same compounds exist in nature, but in lower concentrations.

Methyl mercaptan (CH_3SH) and ethyl mercaptan (CH_3CH_2SH) are often associated with the wood-pulping industry and are troublesome because they can be smelled in concentrations below 1 ppb. They can often be smelled in situations where the concentration is below the detection limit of many analytical methods.

Organic sulfides (R–S–R) and disulfides (R–S–S–R) also exist in air. The lower-molecular-weight (methyl, ethyl, etc.) forms have obnoxious odors.

8.4.7   LONG-LIVED ANTHROPOGENIC COMPOUNDS

Besides those already mentioned, other organic materials are produced by human activity and injected into the air. Some of these have gained notoriety in one way or another as pollutants, and others seem less harmful at this time. The following paragraphs include several classes of compounds that exist in air but have not been thoroughly studied.

Pesticides and their residues are apparently present in all elements of the environment including the atmosphere. These substances are injected directly in spraying or dusting operations, or they may be released with wind-blown dust or by volatilization. Almost no atmospheric data exist on this subject.

Metal organic compounds such as dimethyl mercury or tetraethyl lead can be present as gases in low concentration in air. Direct sources exist for the organo-lead compounds, both in certain industrial process and in the volatilization of leaded fuels. The sources of the organo-mercury compounds are not as well known and may be biological.

Freons such as Freon-12 (CCl_2F_2) are used in large quantity in refrigeration and pressurized spray containers. Much of the industrial production of such inert, long-lived compounds ends up in the atmosphere and may have a very long life there.

8.5 Analytical Methods

8.5.1   CO AND CO_2—NONDISPERSIVE INFRARED (NDIR)

Both carbon monoxide and carbon dioxide exhibit infrared absorption of sufficient strength to be detected over reasonable path lengths at usual urban atmospheric concentrations. In the case of CO, the method is not applicable in background locations due to the low background level (~ 0.1 ppm). The high background of CO_2 makes the method useful in all circumstances.

The infrared spectra of these molecules are unique to them and are shown in Fig. 8.6. The NDIR apparatus is shown in Fig. 8.7. Infrared light of a broad spectrum emanates from the sources and passes through the filter cells, where unwanted wavelength bands are removed (this will

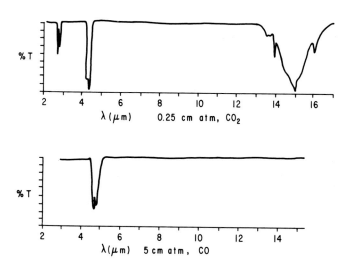

FIG. 8.6. The infrared absorption spectra of CO and CO_2.

be discussed later). Then, the light traverses the reference and sample cells; the former usually contains dry, pure nitrogen, and atmospheric air flows continuously through the sample cell. The chopper alternately turns both beams on and off at a fixed rate, typically 10 Hz.

The key to the NDIR method is the sensor, which is a two-sided cell filled with the gas in question (CO or CO_2). A capillary tube maintains equal static pressure in the two halves of the cell. The cell is divided by an electrically conducting membrane (aluminized plastic, for instance). If there is no CO in the sample cell and if the detector is filled with CO, both sides of the detector will receive a fixed amount of IR light which (due to the chopper) pulsates at a fixed frequency. The membrane will be displaced due to uneven heating toward the cooler side at this same frequency. A capacitance detector senses the movement of the membrane and converts it to an electrical output.

If CO appears in the sample cell, IR light will be removed in the absorption bands of CO, thus making it unavailable for heating the sample side of the detector. This displacement of the membrane will then change as a result of the decrease in heating, and will be sensed electronically.

By adjusting the instrument first with a zero gas (no CO) and with standard gases (e.g., 10 ppm CO or whatever range is desired), it is possible to calibrate the sensitivity of the capacitance detector and associated electronics in units of CO content (or CO_2 if the sensor were filled with CO_2).

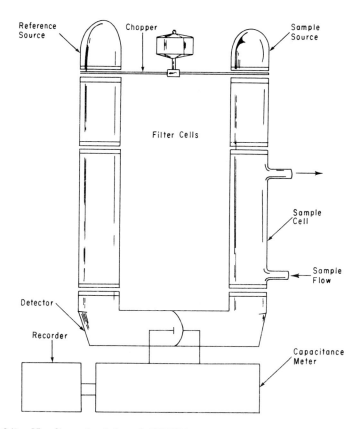

FIG. 8.7. Nondispersive infrared (NDIR) sensor, as used for CO and CO$_2$ measurement.

Due to the fact that there are three major IR absorbing gases in air (CO, CO$_2$, and H$_2$O) and due to their variation (especially that of H$_2$O), interferences exist whenever the bands of one molecule overlap the bands of the one being sensed. The filter cells are designed to compensate for this by removing most of the radiation in the bands of the interfering molecule, so that there is no sensitivity at all in those spectral regions. For the measurement of CO, for instance, the filter cells would contain CO$_2$ and H$_2$O vapor in sufficient concentration. Since no interference can ever occur in the reference cell, some commercial versions of the device use only one filter cell on the sample side.

Virtually all urban CO data in the U. S. are taken with this kind of device. The global CO$_2$ measurements mentioned earlier in this chapter were taken with a version of the NDIR which also has automatic cycling of atmospheric air and standard gases to minimize the effects of drift. The precision so achieved for CO_2 is around ± 0.1 ppm, while the normal urban CO measurements have an uncertainty of about 1 ppm. Some CO instruments exist without filter cells, which can lead to errors as large as ± 5 ppm due to humidity variations. Of course, the accuracy depends on the standard gases used for calibration.

8.5.2 CO—MERCURIC OXIDE METHOD

Hot mercuric oxide reacts with CO to produce CO_2 and mercury vapor:

$$CO + HgO \xrightarrow[210°C]{} CO_2 + Hg$$

The mercury vapor released by this reaction is subsequently sensed by its strong absorption of radiation from a mercury lamp (Robbins *et al.*, 1968).

This method is good over a range of $0.025 \leq [CO] \leq 10$ ppm and is hence useful in background situations. There are interferences from hydrogen and some organic compounds; however, means have been developed to selectively remove such interferences.

8.5.3 HYDROCARBONS—FLAME IONIZATION DETECTION (FID)

Most organic compounds burn in a hydrogen flame to form CO_2 and water vapor with a simultaneous increase in the concentration of gaseous ions in the flame. A sensitive current detector (an electrometer) can be used to sense this increased ion density, providing a measure of the amount of carbon-containing material consumed.

In general, the instrument response is approximately proportional to the number of organically bound carbon atoms released in the flame. Hence, the results are not specific and give only a gross measurement of organic material in units of concentration of the calibration gas, which is usually methane.

Practically all routine hydrocarbon monitoring in the U. S. today is done with the flame ionization detector. The detection limit of the device is around 0.2–0.5 ppm hydrocarbon as methane.

8.5.4 CO, HYDROCARBONS, AND OTHER GASES—GAS CHROMATOGRAPHY (GC)

As discussed in Chapter 5, gas chromatography can be used for CO, hydrocarbons, and many organic molecules. The method was discussed earlier and is mentioned here for the sake of completeness. Samples for GC are usually pretreated to remove interfering substances. After the column, CO is converted to CH_4 in a catalytic reactor and detected with a flame ionization detector. Methane is also detected, as are any other substances not removed in sample treatment; however, a properly designed column provides adequate separation of CO and hydrocarbons.

8.5.5 SPECIFIC ORGANIC COMPOUNDS—WET CHEMICAL METHODS

Methods have been developed and tested for a wide variety of organic substances in air. Some of these methods are sufficiently widely used to deserve mention here. However, since the number of specific compounds for which methods exist is large, and since there is an almost endless list of organic substances which are important, for instance in air pollution, none of the methods will be singled out for detailed study. The sample handling and analytical procedures are usually similar to those methods discussed in Chapters 6 and 7. For instance, bubblers are often used for sampling and a colorimetric analysis follows.

If the reader needs more specific information on these or other methods, references to appropriate sources are included.

Formaldehyde is particularly easy to detect with aqueous reagents since it is very soluble in water. HCHO forms a purple color with chromotropic acid, which provides good sensitivity over the range $0.01 \leq [HCHO] \leq 200$ ppm (U. S. Dept. of Health, Education, and Welfare, 1965; Altshuller *et al.*, 1961).

Acrolein can be detected in a number of ways (Mueller *et al.*, 1971). The 4-hexylresorcinol method provides sensitivity from 20 ppb to 10 ppm with a colorimetric analysis (U. S. Dept. of Health, Education, and Welfare, 1965).

Total aliphatic aldehydes are often measured with a 3-methyl-2-benzo-thiazolone hydrazone hydrochloride (MBTH) reagent. Aldehydes react with the MBTH in the presence of ferric chloride in acid solution to form a blue dye (U. S. Dept. of Health, Education, and Welfare, 1965).

Mercaptans can be measured down to ~ 10 ppb with a colorimetric method using N-N-dimethyl-p-phenylene diamine in dilute HNO_3 and $FeCl_3$ solution, as described by Moore *et al.* (1960).

PROBLEMS

1. Calculate the pH of a raindrop at 25°C with 295 ppm CO_2 and 1000 ppm CO_2. Assume Henry's law.

2. Calculate the concentration of CO_2 in a power station plume resulting from the combustion of a coal containing 60% carbon and 6% hydrogen by 20% excess air. Estimate the concentration of CO_2 due to this source in a valley 2 km wide under a 500-m inversion with a 5-mph wind. Use the box model for the dispersion of CO_2.

3. From Table 8.2, calculate the total mass concentration of hydrocarbons. Compare this mass to typical urban aerosol mass concentrations, and the carbon fraction from Table 8.4. Discuss.

4. Starting with an olefinic hydrocarbon, show how reaction with ozone and then water could create a carbonyl compound.

5. Calculate the number of grams of CO_2 and CO in the atmosphere, assuming mixing ratios of 325 ppm and 0.12 ppm, respectively.

REFERENCES

Altshuller, A. P., Miller, D. L., and Sleva, S. F. (1961). *Anal. Chem.* **33,** 621.
Bolin, B. (1970). *Sci. Amer.* **223,** [3] 124.
Bolin, B., and Bischof, W. (1970). *Tellus,* **22,** 431.
Callendar, G. S. (1958). *Tellus* **10,** 243.
Colucci, J. M., and Begeman, C. R. (1969). *Environ. Sci. Technol.* **3,** 41.
Fleagle, R. G., and Businger, J. A. (1963). "An Introduction to Atmospheric Physics." Academic Press, New York.
Hexter, A. C., and Goldsmith, J. R. (1971). *Science* **173,** 576.
Johnson, K. L., Dworetzky, L. H., and Heller, A. N. (1968). *Science* **160,** 67.
Junge, C., Seiler, W., and Warneck, P. (1971). *J. Geophys. Res.* **76,** 2866.
Keeling, C. D. (1971). Private communication.
Keeling, C. D., and Pales, J. C. (1965). Mauna Loa carbon dioxide project. Rep. No. 3. Scripps Inst. of Oceanography, LaJolla, California.
Koyama, T. (1963). *J. Geophys. Res.* **68,** 3971.
Lonneman, W. A., Bellar, T. A., and Altshuller, A. P. (1968). *Environ. Sci. Technol.* **2,** 1017.
Miller, M. S., Friedlander, S. K., and Hidy, G. M. (1972). *J. Colloid Interface Sci.* **39,** 165.
Moore, H., Helweg, H. L., and Graul, R. J. (1960). *Amer. Ind. Hyg. Ass. J.* **21,** 466.
Mueller, P. K. (1971). (Air and Ind. Hyg. Lab., Dept. of Public Health, State of California). Private communication.
Mueller, P. K., Kothny, E. L., Pierce, L. B., Belsky, T., Imada M., and Moore, H. (1971). *Anal. Chem.* **43,** 1R.
NAS-NRC (1966). "Weather and Climate Modification, Problems and Prospects," Vol. II, Publ. 1350. Nat. Acad. of Sci. (U.S.) and the Nat. Res. Council.
Pales, J. C. and Keeling, C. D. (1965). *J. Geophys. Res.* **70,** 6053.
Pressman, J., and Warneck, P. (1970). *J. Atmos. Sci.* **27,** 155.
Rasmussen, R. A. (1969). *Nat. Meeting Air Pollut. Contr. Ass.,* 1969, Paper 69-AP-26.
Rasmussen, R. A., and Went, F. W. (1965). *Proc. Nat. Acad. Sci. U. S.* **53,** 215.
Revelle, R., and Suess, H. E. (1957). *Tellus* **9,** 18.
Ripperton, L. A., White, O., and Jeffries, H. E. (1967). *Proc. Div. Water Air and Waste Chem., Nat. Meeting, 147th Amer. Chem. Soc., Chicago, Illinois, 1967,* pp. 54–56.

Robbins, R. C., Borg, K. M., and Robinson, E. (1968). *J. Air Pollut. Contr. Ass.* **18,** 106.

Robinson, E., and Robbins, R. C. (1968). Sources, abundance and fate of gaseous atmospheric pollutants. Stanford Res. Inst. Rep. to Amer. Petroleum Inst.

SMIC (1971). Inadvertent climate modification. Rep. of the study of man's impact on climate. MIT Press, Cambridge, Massachusetts.

Swinnerton, J. W., Linnenbom, V. J., and Cheek, C. H. (1969). *Environ. Sci. Technol.* **3,** 836.

Tukey, J. W., Chairman, Environmental Pollution Panel (1965). Restoring the quality of our environment. The White House, Washington, D.C.

U.S. Dept. of Health, Education, and Welfare (1965). Selected methods for the measurement of air pollutants. Publ. 999-AP-11. Publ. Health Serv., Cincinnati, Ohio.

U.S. Dept. of Health, Education, and Welfare (1970a). Air quality criteria for carbon monoxide. Doc. No. AP-62.

U.S. Dept. of Health, Education, and Welfare (1970b). Air quality criteria for hydrocarbons. Doc. No. AP-64.

U.S. Dept. of Health, Education, and Welfare (1970c). Air quality criteria for photochemical oxidants. Doc. No. AP-63.

Went, F. W. (1966). *Tellus* **18,** 549.

Westberg, K., Cohen, N., and Wilson, K. W. (1971). *Science* **171,** 1013.

CHAPTER 9

AEROSOLS

9.1 Introduction

Aerosols are mixtures of gases and particles with small settling velocity that exhibit some degree of stability in a gravitational field (Hidy and Brock, 1970a). The atmosphere is thus an aerosol since it *always* has particulate matter within it. A renaissance occurred in the 1960's in studies of atmospheric aerosols which makes it possible to include this chapter. There seem to be two reasons for this renewed interest in aerosol research, which nearly died early in the century after the work of C. T. R. Wilson, Aitkin, Rayleigh, and their contemporaries had made such a fine start in the field. First, physicists and other basic scientists have been preoccupied for the intervening 50 years or so with atomic and nuclear problems and have only recently returned to more classical physical problems, including aerosol studies. Second, much work on aerosols was done for military purposes in World Wars I and II and thereafter which was kept classified in the United States, England, and probably elsewhere. Some of this work has been released, and more of the current work is being published in the open literature.

This chapter will include a number of the basic aspects of aerosol science that are useful in atmospheric chemistry studies. If a more sophisticated approach is needed, the reader is referred to one of the reference books on the subject. The standard work is "The Mechanics of Aerosols" by Fuchs (1964). "The Dynamics of Aerocolloidal Systems" (Hidy and Brock, 1970a) provides another view—mainly theoretical—of the mechanical behavior of aerosols. "Aerosol Science" (Davies, 1966) provides a review of a broader variety of topics related to aerosol research.

In this chapter, we will first consider the sources and sinks of atmospheric aerosols, at least as far as these are known. Next, we will examine the dominant role of particle size distribution in the behavior of atmospheric aerosols. Then, we will proceed through an elementary view of aerosol mechanics and optics. Finally, we will briefly look at measurement of aerosols, concentrating on *in situ* methods.

9.2 Sources and Sinks of Particulate Matter

9.2.1 Sources

Aerosols are formed by a variety of processes, which can be easily described according to the generation mechanism:

1. *Grinding, impaction,* and other *comminution* processes usually produce dusts in particle sizes that are considered large when such materials are suspended in air. These processes often seem to produce a log-normal size distribution, which will be discussed later. Natural processes such as entrainment of particles by wind (dust storms, etc.) and a variety of industrial operations are examples.

2. *Breakup of liquids,* such as when a bubble bursts at the sea surface, in any sort of industrial spray operation, when a rubber tire passes over wet pavement, etc., produces a *mist* of droplets. These droplets can evaporate, leaving behind any nonvolatile solute present in the original drop.

3. *Condensation* of particles in flames results in the generation of aerosols which often have very small particle size in comparison to the first two processes. Carbon, for example, condenses to form soot particles if a flame has insufficient O_2 or the temperature is too low. Depending on the flame conditions, the condensed particles may agglomerate (or coagulate) to much larger particle size very soon after initial generation. Thus, such thermal sources may produce single particles of largely submicron size. If conditions are different, coagulated clumps or chains of such primary particles may be emitted from the source.

4. *Coagulation* (or agglomeration) of very small particles in the atmosphere may appear to generate larger particles, especially if the sensing instrument is insensitive to the small particles. Coagulation rates will be discussed briefly later in this chapter.

5. *Nucleation* processes can be either homogeneous (where no sizeable condensation nucleus exists) or heterogeneous (where condensation occurs on the surface of an existing particle). Homogeneous nucleation requires very large ratios of vapor pressure to saturation vapor pressure, typically above two (Hidy and Brock, 1970a). If this supersaturation does not exist, stable particles cannot exist for long periods of time, although temporary embryos can. The importance of homogeneous nucleation in the atmosphere is unknown at this juncture, but is probably important in only a few rare instances, possibly in the nearly dust-free upper atmosphere, or at points of emission to the atmosphere where the supersaturation is extremely high.

By comparison, *heterogeneous nucleation* is a well-known process in the atmosphere. In a sense, heterogeneous nucleation competes with a large advantage over homogeneous nucleation because it requires much lower supersaturation. Saturation ratios of 1.05 or even less may result in the formation of stable particles. Since the atmosphere always has particles present to act as nuclei for heterogeneous nucleation, this process is very common. The affinity of the condensate for the surface of the nucleus is an important variable.

The best-studied example (but by no means the only one) is condensation of H_2O on water-soluble nuclei to form stable clouds (which are aerosols). Figure 9.1 shows the calculated saturation ratio versus particle size for water droplets. These curves are based on a combination of Raoult's law and the Kelvin relation for the dependence of vapor pressure of a particle on its radius [see also Section 2.3 and Fletcher (1962)]:

$$\frac{P_{H_2O}}{P^{\circ}_{H_2O}} = \left(\exp \frac{2\sigma \bar{V}}{RTr} \right) \left[1 + \frac{imM_w}{M_s(\frac{4}{3}\pi r^3 \rho - m)} \right]^{-1} \tag{9.1}$$

where σ is the surface tension of a solution droplet in ergs/cm^2; $\bar{V}$ is the molar volume of the liquid phase (volume per mole); T is the temperature in K; r is the droplet radius; i is the van't Hoff factor, the average number of moles of dissolved species produced per mole solute; m is the mass of solute in the droplet; M_s is the molecular weight of solute; ρ is the density of a droplet; and M_w is the molecular weight of water.

Curves based on this equation were first discussed by Köhler (1921) as the basis for droplet growth in a supersaturated gas, and are today referred to as Köhler curves.

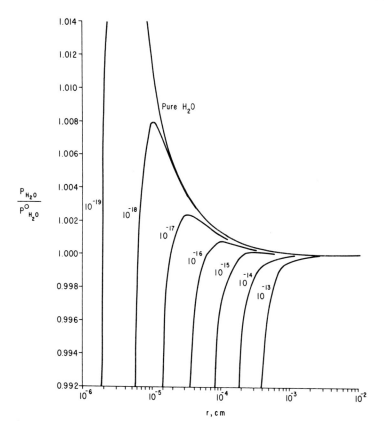

FIG. 9.1. Köhler curves calculated from Eq. (9.1) for the saturation ratio $P_{H_2O}/P^\circ_{H_2O}$ of a water droplet as a function of droplet radius r. The quantity im/M is given as a parameter.

Examples where nucleation processes occur are easy to find in the atmosphere. Besides the obvious case of H_2O clouds, there are other condensed materials found on aerosol particles in air. Organic materials were mentioned in Chapter 8, and H_2SO_4 in Chapter 6. Equation (9.1) could be altered for other substances than water by replacing the properties of H_2O with those of the appropriate volatile substance. Unfortunately, there are few cases where enough is known about condensates other than H_2O in air to permit this.

6. *Reaction on particle surfaces* can occur, causing particle growth. The oxidation of SO_2 may occur on the surface of particles which contain substances acting as catalysts. The resultant sulfate compounds presumably

remain on the surface thereafter. Some substances in fly-ash emissions from coal and oil combustion have been mentioned as promotors of dry SO_2 oxidation.

7. *Droplet reactions* are another class of processes which occur when gaseous materials are absorbed into H_2O droplets in clouds or haze at high humidity followed by reaction and evaporation of the H_2O. An example of this sort of process is the Scott–Hobbs SO_2 oxidation mechanism described in Chapter 6. While it is difficult to realistically describe this process mathematically, the probability of this sort of aerosol generation actually occurring seems high, both on theoretical and observational grounds. Since the mass of material (e.g., SO_2) in the gas phase is usually large in comparison to that in the resulting aerosol particles, this sort of gas-to-particle conversion seems especially important.

8. *Breakup of large particles* can occur mechanically or during phase changes; Radke (1970) has showed that the evaporation of cloud droplets increases the total number of particles which can act as cloud condensation nuclei at small supersaturation. That is, more active particles come out of a cloud—such as a wave cloud—than went into it. It seems possible that the efflorescence process might well result in the breaking of condensed material if it occurred at a fast rate.

Returning to Fig. 1.1a, we can relate this list of processes to those shown in the system diagram. The first three processes are direct sources which introduce material into the air, process (a) in the figure. Coagulation is not a source as such and cannot of itself change the mass of particulate matter in air. Nucleation processes convert gases to particles as in process (c) in Fig. 1.1a. Reactions on dry particle surfaces also come under (c). Droplet reactions are involved with the complex set of processes (d)–(e).

9.2.2 SINKS

It remains to consider the sink processes, which can be listed in a fashion similar to the sources. The approach here is qualitative; those topics that are amenable to simple mathematical description will be expanded later in the chapter.

1. *Settling*, or *sedimentation*, is an important removal process for atmospheric aerosol particles, especially large ones. When such removal occurs, it is referred to as *fallout* or *dry fallout*. The settling velocity is discussed later.

2. *Impaction* of particles is also important for the largest atmospheric particles and occurs when air flow around an object has sufficient curvature to allow the particle inertia to cause it to impact on the object. Removal by large objects such as trees or smaller ones like grass or pine needles provides specific cases.

Impaction also occurs when aerosol particles are collected by falling rain drops. This process—which is efficient for only the largest of atmospheric particles—occurs when the particle has sufficient inertia to not be able to follow the curved lines of flow around the falling drop.

3. ´ *Cloud processes* may or may not remove particles from the atmosphere, depending on the occurrence of precipitation. Both rain and snow remove materials contained in them from the atmosphere.

Dry fallout and impaction are found as process (b) in Fig. 1.1a, and wet removal as process (b), (m), and (q).

9.2.3 GLOBAL AND URBAN CONSIDERATIONS

While it is easy to discuss the sources and sinks of particulate matter in general terms, it is difficult to assemble global or regional budgets for atmospheric aerosol. One of the reasons for this difficulty is the chemical heterogeneity of the particulate matter. Another is the small magnitude of aerosol mass concentrations in air compared to trace gas concentrations and the amounts processed through the atmospheric portion of the elemental cycles that are involved.

Hidy and Brock (1970b) and Robinson and Robbins (1971) have developed global estimates for total particulate mass concentration. Primary particle production is contrasted with secondary particulate matter production by gas-to-particle conversion, and natural aerosols are compared to anthropogenic ones in Table 9.1, adapted from Robinson and Robbins (1971). Thus perhaps 10% or so of the mass of aerosol produced on a global basis is of human origin. Hidy and Brock (1970b) suggested 6%, with a projected increase to perhaps 15% for the anthropogenic portion by 2000 A.D.

While global aerosol production *seems* to be dominated by natural sources, it is safe to conclude on the basis of Table 9.1 that urban aerosols are totally dominated by anthropogenic material. Two-thirds of the area of earth is water, another major percentage is polar, leaving only a small percentage that man can inhabit. Our disposition toward living in cities suggests that the natural aerosols, on the average, should constitute only a very small part, say a few per cent, of the urban particulate matter.

TABLE 9.1

GLOBAL SUMMARY OF SOURCE STRENGTHS FOR
ATMOSPHERIC PARTICULATE MATTER

Source	Strength (Tg/yr)	
	Natural	Anthro-pogenic
Primary particle production:		
Fly ash from coal	—	36
Iron and steel industry emissions	—	9
Non-fossil fuels (wood, mill wastes)	—	8
Petroleum combustion	—	2
Incineration	—	4
Agricultural emission	—	10
Cement manufacture	—	7
Miscellaneous	—	16
Sea salt	1000	—
Soil dust	200	—
Volcanic particles	4	—
Forest fires	3	—
Subtotal	1207	92
Gas-to-particle conversion:		
Sulfate from H_2S	204	—
Sulfate from SO_2	—	147
Nitrate from NO_x	432	30
Ammonium from NH_3	269	—
Organic aerosol from terpenes hydrocarbons, etc.	200	27
Subtotal	1105	204
Total	2312	296

This argument is based entirely on source strengths; that is, we assume an atmospheric concentration proportional to source strength, which is probably not entirely true. There are very few data on *removal* rates or on actual air composition (globally averaged) to improve these estimates. Dust fall rates, which include rainout, washout, and fallout, are measured in many cities. Typical rates of removal are 0.35–3.5 mg/cm^2-month. In very dusty locations, 70 mg/cm^2-month has been reported. Few, if any such data exist from which global budgets could be estimated.

Junge (1963) suggests that precipitation processes account for perhaps 80–90% of the removal of particulate matter from air, with only 10–20% of the removal due to dry deposition. There are some situations where dry removal is more important, for instance, at the sea coast, where large sea-salt particles can impact on vegetation. One of the obvious consequences of wet removal is the modification of hydrometeor composition. Consider-able amounts of work have been done on precipitation chemistry by Junge (1963), Lodge *et al.* (1968), and others showing that rain chemistry is affected—sometimes in important ways—by human activity.

While it is not possible in this text to proceed to much greater length about the composition of rain water, it seems sufficiently important to in-clude Fig. 9.2 [adapted from Odén (1969)] depicting the distribution of pH of rain in Northern Europe. While there is a controversy about the causes, there can be little doubt that there exists a source of acidic material somewhere in the area. The pH of rain in equilibrium with CO_2 is around 5.7, so that the $[H^+]$ maximum concentration in rain is over 10 times the amount expected from the presence of CO_2. Possible sources of H^+ include SO_2, H_2S from anaerobic processes in the Baltic and North Seas, and pos-sibly other sources. The work of Rodhe *et al.* (1971) and Bolin (1971) suggests that SO_2 is the primary source, based on meteorological trajectories and the regional budget of sulfur. Long-distance transport, perhaps more than 1000 km, is suggested by their studies. The impact of this apparent removal of an air pollutant from the atmosphere is mainly on the hydro-sphere. The acidic water affects soil chemistry, fish populations, forest-soil nutrient balances and tree growth, corrosion of metals, decay of calcarious building stones, and so on. It seems safe to suggest that this kind of problem

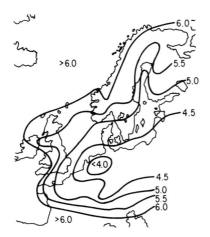

FIG. 9.2. Iso-pH contours for North European precipitation in 1962 (Odén, 1969).

is real, and that one of the future areas for development will be that of precipitation chemistry. Studies of the explicit causes, the relationships of air composition to precipitation effects, and of the proper kind and amount of monitoring are needed. Figure 4.10 showed a map of Cl⁻ ion for the U. S. which suggested that not all substances are grossly changed by human activity. Chloride in precipitation seems to be controlled by the sea.

9.3 Particle Size Distribution

Of all the properties of particles in air, none dominates the behavior more (nor is so difficult to study experimentally) than particle size. The atmospheric aerosol is composed of particles which range in size from large molecules (tens of Ångstroms) up to tens or hundreds of microns. In cgs units, this range is from 10^{-7} cm up to perhaps 10^{-2} cm. No one measurement technique may be used over much more than a decade of size, precluding measurement of a distribution by a single technique.

This chapter will present an approach to the description of particle size by way of examples. Following that, mechanical properties and optical properties will be discussed, especially in their relationship to the all-dominant particle size. Size itself will be represented in two ways—by radius r, and particle diameter D_p. Both terms are used by different workers in the aerosol field today.

Aerosols are not generally presented to us as particles of a single size, shape, and chemical composition, but rather as a distribution of sizes and shapes and a variety of chemical compositions. Since it is difficult to generalize about shape distributions and very little is known about the chemical composition, we shall concern ourselves with a population of spherical particles of uniform density.

Although atmospheric aerosols are too small to be separated with a set of sieve pans, it serves our purpose here to carry out a "thought experiment" in which a set of sieve pans is used to separate a sample of particles. Let us take for our distribution that which results when a large number of particles is shaken in a stack of sieve pans arranged in such a way that the smallest particles settle to the bottom; the largest particles are caught by the topmost pan and those of intermediate size are distributed in the middle pans. We may use an arbitrary number of pans in this experiment. (See Fig. 9.3.)

We may characterize the particles in the sieve pans by their radii, which could be measured quite easily, or by any other single-valued parameter of the particles, including the logarithms of the radii, ln r. The distributions

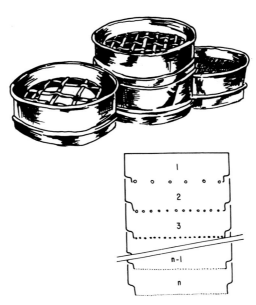

Fig. 9.3. A hypothetical set of sieves for sorting a sample of particulate matter by size. The smallest particles settle to the bottom of the set of sieves, the largest particles stay in the top sieve.

most commonly used are expressed in terms of particle radius or $\ln r$, although there is no reason not to characterize the particles by the mass, volume, or surface area.

The distribution of a number of properties of our sample may be investigated. In what follows, we shall consider the distribution of number of particles, mass, surface, and, for completeness, length in the various sieve pans. Experiments will be described to measure each of these four distributions. Since each of the four distributions may be expressed in term of either r or $\ln r$, there are eight possible distribution functions.

The number distribution could be easily, if laboriously, determined by counting the number of particles in each pan and dividing this number by the total number of particles. (For convenience, all of the distributions we are going to consider will be expressed as *fractions*.) The fraction of particles in each pan shall be denoted by $f(r)\,dr$. The fraction obtained from an experiment in which the particles were characterized by $\ln r$ we shall call $F(\ln r)\,d\ln r$. We might wish to represent $f(r)$ graphically as a continuous function of r as shown in Fig. 9.4. It should be noted that since $f(r)\,dr$ is a fraction, $f(r)$ is a fraction per unit increment of r and has the dimensions r^{-1}. It is important to specify the units of $f(r)$. The fraction of particles with radius between r_1 and r_2 is represented by the area of the rectangle shown in Fig. 9.4 which has the dimensions $f(r)$ by dr. The increment of *number* of particles, dN, in the interval dr is given by $Nf(r)\,dr$, where N is the total number of particles in the sample. The number concentration distribution for an atmospheric problem will be similar to the fraction distribution except that the ordinate might have the dimensions of particles/ cm^4. The distribution function in terms of $\ln r$, $F\,(\ln r)$, will be dimensionless; however, it should be noted that it represents a fraction per unit increment $\ln r$ and not per unit increment $\log r$. All distribution functions we shall discuss will be normalized, that is, $\int_0^\infty f(r)\,dr = 1$, and $\int_0^\infty F(\ln r)\,d\ln r = 1$. The limits of integration will always be given in terms of r.

The mass distribution for the same sample may be very easily determined by weighing the particles in each pan and dividing these weights by the total weight of the sample. The result is once more a fraction which we shall call $f_m(r)\,dr$ or $F_m(\ln r)\,d\ln r$. Since we may choose our set of sieve pans in such a way that the interval dr is arbitrarily small with respect to r, and since our particles are all spheres with uniform densities, the mass distribution function is related to the number distribution by the following equation:

$$f_m(r) = r^3 f(r) \bigg/ \int_0^\infty r^3 f(r)\,dr \qquad (9.2)$$

$f_m(r)$ is also equal to the volume distribution function $f_v(r)$.

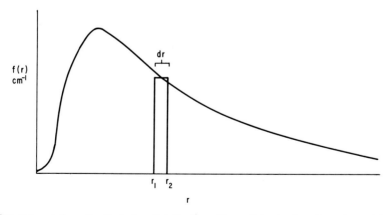

FIG. 9.4. A hypothetical size distribution. Here, $f(r)$, the fraction of particles per increment of radius, is plotted as a continuous function of particle radius r. The fraction of particles between two nearly equal radii r_1 and r_2 is $f(r)\,dr$ and is given by the area of the rectangle.

We also have

$$F_m(\ln r) = r^3 F(\ln r) \bigg/ \int_0^\infty r^3 F(\ln r) \, d\ln r \qquad (9.3)$$

The surface distribution function could be obtained by measuring the total particle surface within each pan (perhaps by a gas adsorption experiment) and dividing this quantity by the total surface for all particles. This fraction $f_s(r) \, dr$ is related to the number distribution function by

$$f_s(r) = r^2 f(r) \bigg/ \int_0^\infty r^2 f(r) \, dr \qquad (9.4)$$

For completeness, we shall also consider the length distribution function, $f_l(r)$. This function could be determined by lining up all the particles in a given pan in a straight line so their edges are just touching and measuring the total length spanned. This length divided by the total length obtained for all pans is the length fraction, $f_l(r) \, dr$. The length fraction is related to the number fraction by

$$f_l(r) = r f(r) \bigg/ \int_0^\infty r f(r) \, dr \qquad (9.5)$$

Finally, without saying how we are going to measure the distribution, the distribution of the qth moment of r may be defined by

$$f_q(r) \equiv r^q f(r) \bigg/ \int_0^\infty r^q f(r) \, dr \qquad (9.6)$$

where q may have any real value.

The average value, or arithmetic mean, of a given particle parameter x is denoted by $\bar{x}$, and given by the expression

$$\bar{x} \equiv \int_0^\infty x f(r) \, dr \qquad (9.7)$$

The average value may be evaluated for any of the distribution functions described here by simply substituting $f_m(r)$, $f_s(r)$, or $f_l(r)$ for $f(r)$. It should be noted that the average value will depend on the distribution function. The average value will be independent of whether the distribution is expressed in terms of r or $\ln r$.

The arithmetic mean of the logarithm of x may also be evaluated for any of the distribution functions:

$$\overline{\ln x} = \int_0^\infty \ln x f(r) \, dr$$

The arithmetic mean of $\ln x$ has a rather special place in the statistics of small particles because this quantity is also the logarithm of the *geometric mean* of x, x_g, defined by the expression

$$x_g = x_1^{f_1} x_2^{f_2} x_3^{f_3} \cdots x_n^{f_n}$$

where f_1 is the probability of observing x_1, etc. It may be seen that these two equations lead to the following equation when x is a continuous function of r:

$$\ln x_g = \int_0^\infty \ln x f(r) \, dr \tag{9.8}$$

Having established some principles and the terminology, we shall now investigate some averages evaluated for a number distribution which is log-normal. Although not all particle samples will be log-normally distributed, this function is observed often enough to warrant its further investigation. The normalized log-normal function is

$$F(\ln r) = \frac{1}{(2\pi)^{1/2} \ln \sigma_g} \exp\left[-\frac{(\ln r - \ln r_g)^2}{2 \ln^2 \sigma_g} \right] \tag{9.9}$$

This function results when the logarithm of the variable r is normally distributed. That this function is normalized may be seen from consideration of the definite integral

$$\int_0^\infty \exp(-z^2/2s^2) \, dz = (\tfrac{1}{2}\pi)^{1/2} s$$

The two parameters required to specify this function are σ_g, the geometric standard deviation, and r_g, the geometric mean radius. These descriptions of r_g and σ_g follow from Eq. (9.8), which defines r_g, and from the following definition of σ_g:

$$\ln^2 \sigma_g = \int_0^\infty (\ln r - \ln r_g)^2 F(\ln r) \, d \ln r$$

Although it is possible to consider many different averages of particle parameters, we shall investigate only a few of the more significant ones.

Many of the other averages, such as average mass, average surface, etc., may be derived directly from the relationships which will be derived here.

The geometric mean radius evaluated from the qth distribution is defined by*

$$\ln r_{gq} = \int_0^\infty (\ln r) F_q \, d \ln r$$

$$= \int_0^\infty (\ln r) r^q F \, d \ln r \Big/ \int_0^\infty r^q F \, d \ln r \qquad (9.10)$$

Let us investigate the term $r^q F$. This may be written

$$\frac{1}{(2\pi)^{1/2} \ln \sigma_g} \exp(q \ln r) \exp\left[- \frac{\ln^2 r - 2 \ln r \ln r_g + \ln^2 r_g}{2 \ln^2 \sigma_g} \right]$$

We may combine the two exponential terms, obtaining after collection of terms in $\ln r$:

$$\frac{1}{(2\pi)^{1/2} \ln \sigma_g} \exp\left[- \frac{\ln^2 r - (2 \ln r)(\ln r_g + q \ln^2 \sigma_g) + \ln^2 r_g}{2 \ln^2 \sigma_g} \right]$$

The square in the exponential may be completed by multiplying and dividing by

$$\exp[(2q \ln r_g \ln^2 \sigma_g + q^2 \ln^4 \sigma_g) / 2 \ln^2 \sigma_g]$$

We now obtain

$$r^q F = \frac{1}{(2\pi)^{1/2} \ln \sigma_g} \exp(q \ln r_g + \tfrac{1}{2} q^2 \ln^2 \sigma_g)$$

$$\cdot \exp\left\{ - \frac{[\ln r - (\ln r_g + q \ln^2 \sigma_g)]^2}{2 \ln^2 \sigma_g} \right\} \qquad (9.11)$$

The geometric mean may now be written

$$\ln r_{gq} = \frac{1}{(2\pi)^{1/2} \ln \sigma_g} \int_0^\infty (\ln r) \exp\left[- \frac{[\ln r - (\ln r_g + q \ln^2 \sigma_g)]^2}{2 \ln^2 \sigma_g} \right] d \ln r$$

$$= \ln r_g + q \ln^2 \sigma_g \qquad (9.12)$$

*For simplicity and clarity, we are replacing $F_q(\ln r)$ with F_q, $F(\ln r)$ with F, etc. in the following discussion.

by direct evaluation of the integral. We not only have a simple relationship between the geometric mean from the number distribution and that obtained from the qth distribution; more significantly, we note from Eq. (9.11) that if the number distribution is log-normal, then the distribution of the qth moment is also log-normal. The two distributions have the same geometric standard deviation, but different values for the geometric mean radius. This property of the log-normal distribution was first recognized by Kapteyn [see, for example, Herdan (1960)]. The relationships connecting the number, length, surface, and mass distribution functions for a log-normal distribution are depicted in Fig. 9.5a for a case where $\ln \sigma_g = 1.7$. The logarithmic probability plot (Fig. 9.5b) makes use of the fact that $\sigma_g \simeq r_{0.84}/r_g$, where $r_{0.84}$ is the radius larger than that of 84% of the particles in the sample.

The arithmetic mean radius evaluated from the qth distribution is

$$r_q = \int_0^\infty r F_q \, d \ln r$$

$$= \int_0^\infty r^{q+1} F \, d \ln r \Big/ \int_0^\infty r^q F \, d \ln r = R_{q+1}/R_q \qquad (9.13)$$

It can be shown by direct integration that the integral R_q is given by the following expression:

$$R_q \equiv \int_0^\infty r^q F \, d \ln r = \exp[q \ln r_g + \tfrac{1}{2} q^2 \ln^2 \sigma_g] \qquad (9.14)$$

from which we may obtain

$$\ln r_q = \ln r_g + (q + \tfrac{1}{2}) \ln^2 \sigma_g \qquad (9.15)$$

The arithmetic mean of the pth power of r evaluated from the qth distribution, $(r^p)_q$, is equal to R_{p+q}/R_q, or

$$\ln (r^p)_q = p \ln r_g + \tfrac{1}{2} p (p + 2q) \ln^2 \sigma_g \qquad (9.16)$$

Finally, the geometric mean of r^p evaluated from the qth distribution is given by

$$\ln (r^p)_{gq} = p \ln r_{gq} = p \ln r_g + pq \ln^2 \sigma_g \qquad (9.17)$$

Size distributions measured in the atmosphere show many variations, but also show some useful and marked regularities. Perhaps the most complete data were taken in Los Angeles, California, by Whitby et $al.$ (1972a,b) during a large cooperative experiment in 1969. Fewer data have

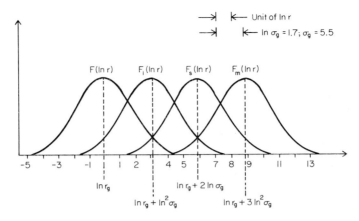

Fig. 9.5a.   A log-normal distribution (ln σ_g = 1.7) and its moments. $F(\ln r)$ is the *number* distribution function, $F_1(\ln r)$ the *length* distribution function, $F_s(\ln r)$ the *surface* distribution function, and $F_m(\ln r)$ the *volume* and *mass* distribution function. The ordinate is the distribution function and the abscissa is ln r.

been taken in remote locations. Figure 9.6 is a typical *number* distribution for urban smog aerosol. Here, $\log(dN/dD_p)$ is plotted versus log D_p, where D_p is the particle diameter. These same data are plotted in another fashion in Fig. 9.7, with several other distributions from various sources. Here, the ordinate is the increment of *volume* per log D_p interval, with the abscissa log D_p. While the area under the curve in Fig. 9.6 has no simple physical meaning (since there is no zero on a log–log plot), it is easy to see that the areas under the curves of Fig. 9.7 are proportional to the volume concentration ($\mu m^3/cm^3$). This is in turn proportional to mass concentration via density. It is interesting to note the almost universal bimodal character of these data, which disappears if plotted by number as in Fig. 9.6. Whitby *et al.* (1972a,b) have suggested that the small-particle mode contains the products of gas-to-particle conversion and coagulation, while the large mode ($r \gg 1~\mu m$) is wind-blown dust, etc.

As discussed earlier, it is useful to be able to describe size distributions (of number, volume, or whatever) by a simple mathematical expression. The simplest description is a two-parameter function in the form of a

Fig. 9.5b.   The same distributions as in Fig. 9.5a plotted as cumulative quantities on a logarithmic probability plot. The ordinate is ln r and the abscissa is the percent of the integral of the function below a given radius. The nonlinear percentage scale is derived from Eq. (4.3). See also Fig. 4.7.

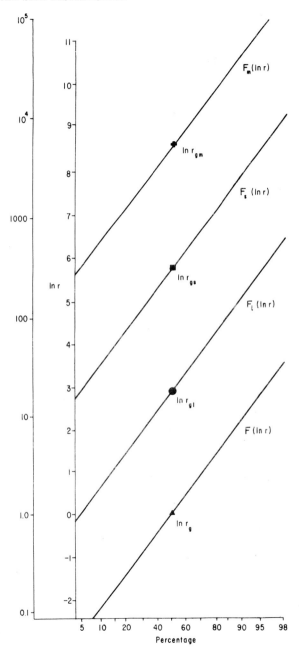

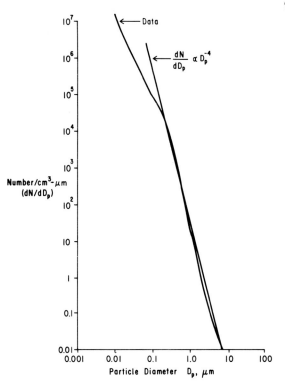

FIG. 9.6.   Grand average *number* distribution in Pasadena, California in August and September of 1969 (Whitby *et al.*, 1972a,b). The number distribution function is plotted as dN/dD_p number cm^{-3} μm^{-1}) versus D_p (μm) on log–log scales. For comparison to a power-law function, a line for $dN/dD_p \propto D_p^{-4}$ is also given. Note that the area under the curve carries no simple meaning.

power law as sugested by Junge (1963):

$$F(\log r) = dN/d(\log r) = Cr^{-\beta} \qquad \text{or} \qquad dN/dD_p = C'D_p^{-(\beta+1)} \quad (9.18)$$

etc., where C and C' are functions of concentration and β describes the relative amounts of large and small particles. The data in Fig. 9.6 can indeed be approximated by this simple expression, with $dN/dD_p \propto D_p^{-4}$, or $\beta \simeq 3$. Besides the variables C (or C') and β, it is necessary to establish the lower size limit.

The next, more complex description is the *log-normal* function, which has three parameters: concentration, r_g, and σ_g. Figure 9.8 shows the comparison of the average Los Angeles distribution from Figs. 9.6 and 9.7 with a log-normal distribution of geometric mean diameter (by volume) of 0.302 μm and $\sigma_g = 2.24$.

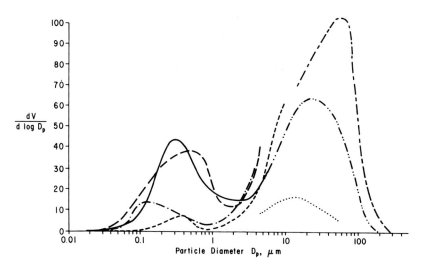

FIG. 9.7. A comparison by Whitby *et al.* (1972a) of several *volume* distribution functions, including the data from Fig. 9.6. Here, the volume increment per log size interval, $dV/d(\log D_p)$, is plotted on a *linear* axis versus size on a *log* axis. The area under the curve represents the total volume. The locations are: (a) (——) Pasadena, California; (b) (— —) and (c) (—·—) Minneapolis, Minnesota; (d) (- - -) Fort Collins, Colorado; (e) (— - —) Seattle, Washington; (f) (—··—) Germany (various locations); (g) (···) Japan.

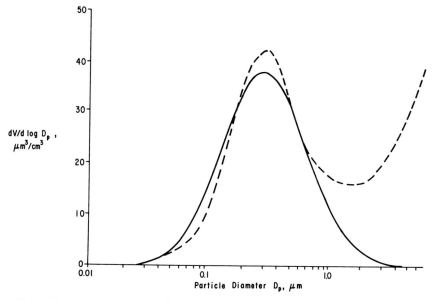

FIG. 9.8. A comparison of the Pasadena, California volume distribution function $dV/d(\log D_p)$ (dashed curve) with a log-normal curve (solid curve). The axes are the same as in Fig. 9.7. $r_{gm} = 0.302 \ \mu m$, $\sigma_g = 2.24$. (Whitby *et al.*, 1972a.)

Other, more complex distribution functions have appeared in the literature, but for our purposes, those considered here are adequate. The more
complex functions contain more parameters which allow a better fit to data.

9.4 Mechanical Properties of Aerosols

9.4.1 DRAG FORCE

The motion of particles in aerosols is dominated by the drag force on
the particle as it moves through the gas. The force is in the direction opposite to the motion of the particle.

Movement due to applied forces is often considered to be balanced by
the drag force, yielding a steady-state condition. In the case where an
imbalance exists, acceleration or deceleration occurs, with the time constant being dictated by the drag force and particle inertia.

Drag force is most easily described for three separate classes of particle
size, based on the relative magnitudes of particle size and mean free path
in the gas. If we use the ratio of mean free path to particle size (l/r) as an
index of this relationship, we can define the three classes as follows:

$l/r \ll 1$ large particles

$l/r \simeq 1$ particle size of the scale of the mean free path

$l/r \gg 1$ small particles

The ratio l/r is known as the Knudsen number, Kn. Since $l \simeq 7 \times 10^{-6}$ cm
in air at STP, the three classes of particle size can be stated quantitatively:

$r \gg 0.07$ μm   Kn $\ll 1$

$r \simeq 0.07$ μm   Kn $\simeq 1$

$r \ll 0.07$ μm   Kn $\gg 1$

Kn $\ll$ 1 particles. Stokes's formula describes the drag force, $F_{\text{drag}-1}$, on
large spherical particles:

$$F_{\text{drag}-1} = -6\pi\eta r V \qquad (9.19)$$

where η is the viscosity of the gas, r is particle radius, and V is velocity.
Several assumptions were necessary in the derivation of this equation: (1)
spherical particles; (2) incompressible medium (i.e., constant density of
gas); (3) infinite extent of medium; (4) low velocity (velocity small with
respect to gas molecule velocity); (5) constant velocity; (6) rigid par-

ticles; (7) no fluid slip at surface of particle (i.e., velocity of gas in immediate contact with particle surface is the same as the particle velocity).

Fuchs (1964) provides a discussion of the basis for these assumptions and Lamb (1945) derives the Stokes drag force from fluid-mechanical considerations.

Kn $\simeq$ 1 particles. As the particle size decreases to a point where $r \simeq l$, the assumption (7) that the fluid in contact with the particle has the same velocity as the particle ceases to be valid. This situation is referred to as "slip" because the drag force decreases below that which would be expected from the Stokes equation:

$$F_{\text{drag-2}} = -6\pi\eta r V (1 + A \cdot \text{Kn})^{-1} \qquad (9.20)$$

Fuchs (1964) and others provide estimates of A which vary slightly with the nature of the surface of the particle:

Aerosol	A
Oil drops	0.87
Rough spheres	0.7
Aqueous solution drop	0.82–0.9

This and the other formulas used to described drag force in the $\text{Kn} \simeq 1$ regime show that F decreases as Kn increases, so A is sometimes called the Cunningham slip correction.

Kn $\gg$ 1 particles. In the case of very small particles (in air at STP, $r \ll 7 \times 10^{-6}$ cm), the particles behave like large molecules whose motion is governed by collisions. The drag force in this regime, $F_{\text{drag-3}}$, has a simple form (Hidy and Brock, 1970a) which can be stated for motion in one direction

$$F_{\text{drag-3}} = (8/3) r^2 N_g (2\pi m_g kT)^{1/2} [1 + (\pi\alpha_m/8)] V \qquad (9.21)$$

where N_g is the number of gas molecules per cubic centimeter of mass m_g, k is Boltzmann's constant, and α_m is an empirical thermal accommodation coefficient for the collision of gas molecules with the particle surface. Since $0 \leq \alpha_m \leq 1$, the effect of α_m is not very great. Also, experimental evidence suggests high values of α_m for air and many surfaces, so that further simplification for practical purposes could be made by simply assuming $\alpha_m \simeq 1$.

This equation is based on many assumptions, the most important of which is that the velocity V is much less than the thermal velocity of the gas molecules. This condition is reasonable for lower-atmosphere aerosol particles settling under the force of gravity.

9.4.2   MOBILITY

For some applications, it is more convenient to use the term mobility B defined as velocity per unit applied force. Since applied force is equal in magnitude to drag force for a steady-state situation,

$$B = V/-F_{drag} \tag{9.22}$$

where

$$B = 1/6\pi\eta r \qquad \text{for} \quad \text{Kn} \ll 1$$

$$B = (1 + A\cdot\text{Kn})/6\pi\eta r = [1 + (Al/r)]/6\pi\eta r \qquad \text{for} \quad \text{Kn} \simeq 1$$

$$B \simeq 3r^{-2}N^{-1}_g(2\pi m_g kT)^{-1/2}/(8 + \pi\alpha_m) \qquad \text{for} \quad \text{Kn} \gg 1$$

Figure 9.9 shows the transition from a $B \propto 1/r$ dependence for large particles to a $B \propto 1/r^2$ dependence for small particles.

9.4.3   SETTLING VELOCITY

Under the force of gravity, particles settle with velocities that depend on their drag force or mobility. Assuming steady-state motion, we can set the gravitational force equal to the drag force F_{drag}. If we have spherical particles of density ρ, this equation can be written:

$$\tfrac{4}{3}\pi r^3(\rho - \rho_{air})g \simeq \tfrac{4}{3}\pi r^3\rho g = -F_{drag}$$

Since $\rho \gg \rho_{air}$, buoyancy is usually unimportant. If we are in the Stokes drag force regime, this becomes

$$V_s = 2r^2\rho g/9\eta \tag{9.23}$$

Other equations can be written for $\text{Kn} \simeq 1$ and $\text{Kn} \gg 1$ drag force regimes.

When actual quantities are put into Eq. (9.23), it is obvious that particles in air at sea level with $r < 1$ μm have such low settling velocities that they can have substantial lifetimes in the atmosphere. For the smallest particle sizes in the atmosphere, it is also clear that removal mechanisms other than settling (or sedimentation) may be dominant.

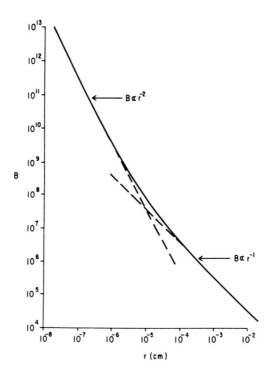

FIG. 9.9.   Mobility B as a function of particle size r in cgs units. The $B \propto r^{-2}$ dependence at $r \ll 10^{-5}$ cm is shown, as is the $B \propto r^{-1}$ dependence when $r \gg 10^{-5}$ cm.

9.4.4 STOKES'S NUMBER

If an aerosol is subjected to flow with curved streamlines, the possibility exists for inertia to cause the particles to separate from the gas. Detailed mathematical solutions for the motion of the gas and of the particles are available in the book by Fuchs (1964) and elsewhere. However, it suffices here to provide a justification for the use of the dimensionless Stokes number (Stk) as an index of the degree of particle–gas separation in curved flow.

The simplest case of inertial separation (albeit hypothetical) would be the situation described in Fig. 9.10.

Assuming that the particles and gas in the aerosol had an initial velocity of V_0, the acceleration experienced by the particle after it passes the hole is given by Newton's Second Law

$$m\, dV/dt = F_{\text{drag}} = -V/B$$

or

$$dV/V = -dt/Bm$$

which integrates to the following if the initial velocity is V_0 at $t = 0$:

$$V = V_0 e^{-t/Bm} = V_0 e^{-t/\tau} \tag{9.24}$$

The quantity $\tau = Bm$ has units of time and is called the inertial period. It is the time required for a particle to decelerate to $1/e$ of its original velocity, and is independent of V_0.

Now, in the same physical situation, the particle will decelerate exponentially, hence only asymptotically approaching zero velocity. But we find that it goes a fixed distance in decelerating to $V = 0$. By integrating the previous equation and setting $V = dx/dt$:

$$\int_0^{\mathcal{L}} dx = V_0 \int_0^{\infty} e^{-t/\tau}\, dt, \qquad \mathcal{L} = V_0\tau = V_0 Bm \tag{9.25}$$

The quantity $\mathcal{L}$ is called the *stop distance* for a particle. The Stokes number Stk can be mathematically derived from considerations of streamline geometry, but is approximately given by

$$\text{Stk} = \mathcal{L}/l \tag{9.26}$$

where l is the radius of curvature of the streamlines.

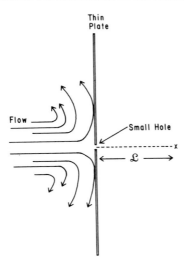

FIG. 9.10. A hypothetical situation for observing the inertial period of an aerosol particle. Fluid containing the particle impinges on a plate with a small hole, and only the particle goes through the hole. The stop distance is $\mathcal{L}$, as described in Eqs. (9.24) and (9.25).

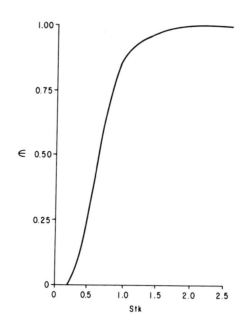

FIG. 9.11. Typical efficiency of an impaction device for removal of particles from a gas stream. Efficiency ϵ is plotted versus Stokes's number Stk as defined in Eq. (9.26).

Aerosol sampling impactors are designed via this approach so that a given impaction stage collects all particles with mB greater than a given given value. Figure 9.11 shows the collection efficiency ϵ for a typical impactor (Fuchs, 1964).

9.5 Diffusion and Coagulation

In 1905, Albert Einstein showed that Brownian motion of particles suspended in a fluid was a necessary result of the molecular-kinetic theory of heat. In so doing, he linked together the mechanical behavior (i.e., the mobility of particles) with the phenomenon of diffusion (Einstein, 1908). The brief discussion below summarizes these conclusions, with special note of the process of coagulation.

The basic statement relating the Fick diffusion coefficient D (as defined in Chapter 1) with particle mobility B is

$$D = kTB \qquad (9.27)$$

where k is Boltzmann's constant. The derivation of this expression is simple, and has been published extensively in physics texts. A particularly interesting version was developed in the paper by Einstein (1908), which has been republished in English (Einstein, 1956). Rewriting Fick's law for diffusion in terms of number concentration N [see Eq. (1.5)], we have

$$dN/dt = kTB \ \nabla^2 N$$

if B and T are not dependent on the spatial coordinates.

Many solutions to these diffusion equations have been published which cover the usual geometries of pipes, containers, and surfaces to which diffusion occurs. There are two particularly important solutions for the atmosphere—diffusion to a flat horizontal surface (including sedimentation) and diffusion to a sphere (such as a cloud drop).

Although it is obviously limiting to assume that the atmosphere is at rest, it is instructive to do so. The number of particles deposited on a horizontal surface of area A between t and $t + dt$ is

$$dN_{dep} = N_0 A \left\{ \left(\frac{D}{\pi t} \right)^{1/2} \left(\exp - \frac{V_s^2 t}{4D} \right) + \frac{V_s}{2} \left[1 + \mathrm{erf} \left(\frac{V_s^2 t}{4D} \right)^{1/2} \right] \right\} dt \quad (9.28)$$

where N_0 is the initial particle concentration in an aerosol assumed to be of infinite extent. It is assumed that all particles reaching the surface are deposited (an accommodation coefficient of unity). Substitution of actual values for a typical atmospheric aerosol in this equation (of course, as a function of size) leads to the conclusion that settling is important for large particles, but diffusion is important for the smallest ones.

The case of diffusion to a sphere of radius R in an infinitely large volume of aerosol of concentration N_0 results in the deposition of N_{dep} particles as a function of time on the whole sphere:

$$N_{dep} = 4\pi R D N_0 [t + 2R(t/\pi D)^{1/2}] \quad (9.29)$$

For very large R or small t, this solution is the same per cm² sec as that for a flat surface without sedimentation. The other extreme, small R and large t, leads to dominance of the first term of the sum, i.e.,

$$N_{dep} \simeq 4\pi R D N_0 t \quad (9.30)$$

with a rate of deposition of

$$dN_{dep}/dt = 4\pi D R N_0 \quad (9.31)$$

This expression has wide application in problems of cloud physics and atmospheric aerosols since R is usually small (e.g., 10^{-3} cm for a cloud drop) and t is large (minutes to hours for the life of an identifiable cloud of particles).

Now, if R approaches the size of the aerosol particles themselves, we can see that we are considering the process of coagulation. Since two spherical particles of radius r will meet when their centers are $2r$ apart, Eq. (9.30) can be rewritten for the number of particles colliding with a stationary "test particle" of the same size (at long time, as before):

$$N_{\text{dep}} = 8\pi DrN_0 t \qquad (9.32)$$

Since there are no "fixed" particles, the relative diffusion coefficient must be used, which is just $2D$ if all particles have the same size. Thus there are $16\pi DrN_0$ collisions per particle per second, or $(\frac{1}{2}N_0)16\pi rDN_0 = 8\pi rDN_0^2$ collisions per second per cm³ of aerosol (the factor of $\frac{1}{2}$ is necessary to avoid counting each collision twice). If the accommodation coefficient is unity, this is the coagulation rate for a monodisperse aerosol of concentration N is

$$dN/dt = -8\pi rDN^2 = -KN^2 \qquad (9.33)$$

where the coagulation constant is given by

$$K = 8\pi rD = 8\pi rkTB$$

Few studies of Brownian diffusion of particles in the atmosphere have ever been made even though it is probably an important mechanism for removal of small particles. Several theoretical studies of coagulation have been made for the atmospheric problem, and a few measurements have been obtained for comparison to the theory.

Junge and Abel (1965) show the calculated change in size distribution to be expected in a hypothetical initial size distribution, as shown in Fig. 9.12. The key features of this calculation are the rapid disappearance of small particles ($r < 0.1$ μm) and the small increase in size of collision partners ($r > 0.1$ μm).

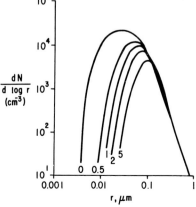

Fig. 9.12. The calculated effect of coagulation on size distribution (Junge and Abel, 1965). The number distribution function $dN/(d \log r)$ (in cm⁻³) is plotted versus particle size r in μm). Values on the curves indicate days. The initial size distribution (assumed) is denoted by zero days and the effects at intervals up to five days are shown.

Husar *et al.* (1972) show the effects of aging atmospheric aerosols in a closed sampling container (plastic bag). The main conclusion is that for Los Angeles aerosol the smallest particles ($D_p < 0.01$ μm) are lost to the particles around $D_p \lesssim 0.1$ μm. Particles in the range around $D_p \gtrsim 1$ μm are not strongly effected by coagulation.

9.6 Optics of Aerosols

The presence of molecules and particulate matter in air results in an interaction with light called *scattering*. This phenomenon is one of the most obvious consequences of atmospheric aerosols, resulting in effects ranging from the very common visibility degradation in cities to the appearance of rare optical occurrences such as the blue moon. Our purpose in this section will be to provide a description of the most important aspects of this problem area, without recourse to detailed mathematical derivation.

9.6.1   RAYLEIGH SCATTERING

Light interacts with particles in a gas (from molecular size on up) by virtue of the electromagnetic wave nature of light and the presence of electrons and positive charges in matter. Particles small relative to the wavelength of light experience the same electric field throughout their entire dimensions. This external electric field establishes a dipole in the small particle that is a function of its polarizability. Since the field varies in time, so does the dipole, which then must emit radiation, as do all oscillating electric dipoles. This emission of light results in the removal of energy from the incident beam and a redirecting of the radiation in the characteristic dipole pattern. Lord Rayleigh established the basic nature of this *scattering* process, and it is known therefore as Rayleigh scattering. The intensity of scattered light I_ϕ from an ideal dielectric sphere is given by (including both planes of polarization)

$$\frac{I_\phi}{E} = \frac{8\pi^4 r^6}{R^2 \lambda^4} \left(\frac{\mathfrak{M}^2 - 1}{\mathfrak{M}^2 + 2}\right)^2 (1 + \cos^2\phi) \tag{9.34}$$

where ϕ is the scattering angle (measured from the transmitted beam), $\mathfrak{M}$ is the refractive index of the scattering *particles*, R is the distance from the point where scattering occurs, λ is the wavelength of light, r is the particle radius, and E is the illuminance (luminous energy per unit area and time). One major consequence of this relationship is that the scattering by gases, including air, has the same λ^{-4} wavelength dependence independent of viewing angle. Rayleigh scattering by the molecules in air is sometimes dominant, especially in clean air. In haze, smoke, smog, etc., the scattering by molecules is usually dominated by Mie scattering by the aerosol particles.

9.6.2 MIE SCATTERING

As the particles become larger relative to the wavelength of light, it is no longer possible to consider the dipole to be exactly in synchronism with the external wave. Rather, it is necessary to consider the wave inside the particle along with the external wave. The theoretical description of this situation was first developed by Mie (1908), and the results are known as the Mie formulas or the Mie theory. Discussions of the Mie theory are also given by Van de Hulst (1957) and Kerker (1969). These relationships describe the process of scattering for particles of any size (including the Rayleigh case). The general character of these solutions for particles where $r \gtrsim \lambda$ is the result of constructive and destructive interference by the internal and external waves, leading to complicated mathematical formulations. Figure 9.13 shows the calculated scattered intensities in both planes of polarization (i_1 and i_2) as functions of scattering angle ϕ for a particle with $r/\lambda \simeq 10$.

If, instead of a single particle, an aggregation of particles of different sizes (such as atmospheric aerosol) is considered, much of the complexity of this diagram disappears. Figure 9.14 shows typical measured intensity in the atmosphere as a function of angle. While the oscillatory nature of Fig. 9.13 is not present, the large amount of scattering at small angles—called forward scattering—clearly dominates in this typical atmospheric case.

It is important to note that Mie scattering often dominates the Rayleigh scattering by the air molecules in haze or smog, so that the angular scattering diagram is very nearly that of the particles alone. In very clean situations, both the gas molecules and aerosol particles contribute to the observed scattering.

9.6.3 EXTINCTION

The scattered intensity is only one of the quantities that needs to be considered in atmospheric light scattering. One of its main consequences is the *extinction* of light from a beam. Examples of extinction in the atmos-

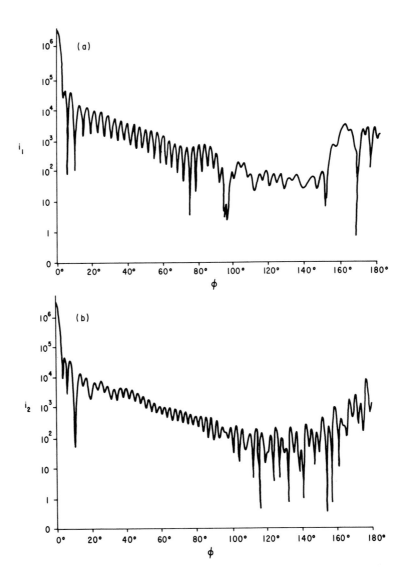

FIG. 9.13.   Scattered intensity in one plane of polarization (i_1) as a function of scattering angle ϕ for $\chi = 60$ ($r/\lambda \simeq 10$). The calculation assumes spherical particles of refractive index, $\mathfrak{N} = 1.5$. i_1 is in arbitrary units (Bullrich, 1964). (b) The same as (a), but for i_2, the other plane of polarization.

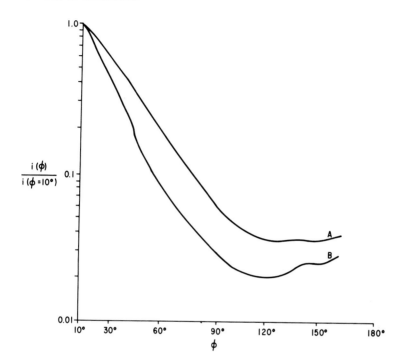

Fig. 9.14.   Measured scattered intensity as a function of angle relative to scattered intensity at 10°, including both planes of polarization using blue light. The ratio $i(\phi)/i(\phi = 10°)$ is plotted versus ϕ for two different times: A, December 19, 1961; and B, November 23, 1961 in Mainz, Germany (Bullrich, 1964).

phere range from the diminution of direct sunlight to the decrease in visibility of a distant bright object. Middleton (1952) relates the volume scattering function $\beta'(\phi)$ to scattered intensity by a volume v of aerosol (including both aerosol—Mie—and gas molecule—Rayleigh—scattering) using the relation

$$I_\phi = E\beta'(\phi)v \qquad (9.35)$$

where E is the illuminance of a collimated beam incident upon the volume of scatterer. All of the angular information is contained in $\beta'(\phi)$, as is the information on the effectiveness of the medium to scatter light. This volume scattering function is further defined when integrated over 4π steradians, since the amount of light removed from the beam (in the process of extinction) must equal the total scattered in all directions:

$$2\pi \int_0^\pi \beta'(\phi) \sin\phi \, d\phi = b_{\text{scat}} \qquad (9.36)$$

where b_{scat} is defined in the Beer–Lambert law as the extinction coefficient in a scattering system [see Eq. (2.16)].

Here, b_{scat} is the sum of scattering by both molecules (Rayleigh scattering) and particles (Mie scattering), which can be represented by

$$b_{scat} = b'_{scat} + b_{Rayleigh}$$

where b'_{scat} represents the aerosol scattering component of extinction and $b_{Rayleigh}$ is the scattering of the gas molecules. It is both customary and convenient to also relate the extinction coefficient to *scattering cross section*. If we consider the extinction coefficient of a monodisperse aerosol, b'_{scat}, we have

$$b'_{scat} = NS \tag{9.37}$$

where N is the number concentration of particles (cm^{-3}). The quantity S thus defined has units cm^2 and is called the scattering cross section. For a polydisperse system of particles, with size classes denoted by the subscript i,

$$b'_{scat} = \sum_i N_i S_i \tag{9.38}$$

For very large objects ($r \gg \lambda$), such as buildings, baseballs, etc., the scattering cross section is just twice the geometric cross section. This seeming paradox—that the particle removes more energy from the beam than it intercepts—is explained by the fact that energy is lost from the beam by interception and *also* by diffraction. If the object (say, a house) is large compared to λ, the diffraction angle is small, so that one would have to be very far away to measure the loss due to diffraction. In the case of $r \gg \lambda$, the amounts lost by interception and diffraction are equal, giving $S = 2 \times$ area.

When we consider the case of $r \simeq \lambda$, then an additional effect occurs due to the disturbance of the electromagnetic field at some distance from the particle. Scattering may occur from the region of the disturbance rather than from just the volume of the particle, leading to $S > 2 \times$ area. It is usual to describe the *scattering efficiency* Q_{scat} of a particle of cross-sectional area A by

$$Q_{scat} = S/A \tag{9.39}$$

which is dimensionless, and has a value of two at $r \gg \lambda$. It is also usual to use a dimensionless size parameter rather than r, namely $\chi = 2\pi r/\lambda$. Figure 9.15 is an example of the calculated dependence of Q_{scat} on χ for two refractive indices. Values of Q_{scat} well above two are observed in these figures, as is the very strong dependence of Q_{scat} on χ when $\chi < 1$, i.e., when $r < \lambda$.

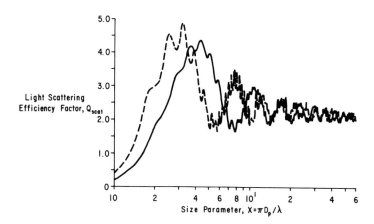

FIG. 9.15. Scattering efficiency Q_{scat} as a function of size parameter χ for refractive indices of (—) 1.50 and (- - -) 1.70 (Ensor et al., 1972).

This strong dependence of Q_{scat} on χ leads to a strong dependence of db'_{scat}/dr on r in the atmosphere. It is most useful to look at the dependence of $db'_{scat}/d(\log D_p)$ on $\log D_p$ (D_p=particle diameter) since a wide range of diameters must be considered. The area under the ogival curve in Figure 9.16 is the scattering coefficient. In this case, the scattering coefficient per log diameter interval was calculated for the aerosol alone from a measured size distribution in Los Angeles, California as a function of wavelength (Ensor et al., 1972). It can be seen that the first peak of the Q_{scat} versus χ curve dominates, yielding the result that particles in the narrow range of sizes $0.2 < D_p < 1$ μm account for almost all of b_{scat} in the case of Los Angeles aerosol. A refractive index of 1.5 was chosen; however, only minor changes would be noticed if $1.3 \leq \mathfrak{M} \leq 1.7$.

In the case of extinction by a particle-free, nonabsorbing gas, a simpler expression holds

$$b_{Rayleigh} = \frac{8\pi^3}{3\lambda^4} \frac{(\mathfrak{M}_g{}^2 - 1)^2}{N_g} \frac{6 + 3f_D}{6 - 7f_D} \tag{9.40}$$

where $\mathfrak{M}_g$ is the refractive index of the gas itself relative to vacuum, N_g is the number of molecules per cm³, and f_D is the depolarization factor, which for air is about 0.042 and is often neglected.

9.6.4 ABSORPTION

Aerosols composed of materials that absorb light (iron oxides, soot, etc.) will absorb light. The mathematical description of this phenomenon is

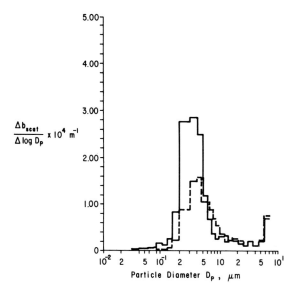

FIG. 9.16.   Light scattering distribution function $\Delta b_{scat}/\Delta(\log D_p)$ on a linear scale versus D_p on a log scale for two wavelengths, $\lambda = 365$ nm (———) and $\lambda = 546$ nm (- - -). The area under the curve is the cattering coefficient b_{scat}. A refractive index of 1.50 was summed int his calculation, which is based on a measured size distribution in Pasadena, California (Ensor et al., 1972).

based on the use of a complex number for describing the refractive index $\mathfrak{M}$. The real part of the complex number is the ordinary refractive index, while the imaginary part respresents absorption. This complex refractive index is used in the Mie formulas to calculate the absorption coefficient.

Little is known about light absorption by atmospheric aerosols, regardless of location. It is usual to assume that the extinction component due to scattering dominates that due to absorption, i.e., $b'_{scat} \gg b_{abs}$. In theoretical work, the approach described here for pure scattering is extended to include absorption. The total extinction coefficient b is the sum

$$b = b_{scat} + b_{abs}$$

which is the sum of integrals (assuming no absorption by gases)

$$b = \pi \int Q_{scat} r^2 \, dN + \pi \int Q_{abs} r^2 \, dN + b_{Rayleigh} \tag{9.41}$$

which defines the absorption efficiency factor Q_{abs}.

9.6.5 WAVELENGTH DEPENDENCE OF EXTINCTION

It has been known since the time of Tyndall that scattering of colloidal materials (hydrosols as well as aerosols) exhibits complex behavior with respect to the wavelength of light. If such a dispersion consists of particles of a single size, the colors can be vivid, with a strong dependence on scattering angle. If we consider a polydisperse aerosol, however, the wavelength dependence of extinction is dictated entirely by the size distribution (assuming no absorption and $b'_{scat} \gg b_{Rayleigh}$).

Using the simplest mathematical representation of size distribution for atmospheric aerosol—the power law—we can derive a relationship for the wavelength dependence of extinction. First, the extinction coefficient can be represented for the aerosol particles alone as an integral, where we denote Q_{scat} as a function of r, λ, and $\mathfrak{M}$, the refractive index:

$$b'_{scat} = \int_0^\infty Q_{scat}(r, \lambda, \mathfrak{M}) \pi r^2 (dN/d \ln r) \, d \ln r \qquad (9.42)$$

and since $dN/d(\ln r) = Cr^{-\beta}$ and $d \ln r = dr/r$,

$$b'_{scat} = \pi C \int_0^\infty Q_{scat}(r, \lambda, \mathfrak{M}) r^{1-\beta} \, dr$$

But r and λ are linked in Q by $\chi = 2\pi r/\lambda$, so $r = \chi\lambda/2\pi$ and $dr = \lambda \, d\chi/2\pi$. Thus,

$$b'_{scat} = \pi C \int_0^\infty Q_{scat}(\chi, \mathfrak{M}) (\chi\lambda/2\pi)^{1-\beta} (\lambda/2\pi) \, d\chi \qquad (9.43)$$

$$= \pi C (\lambda/2\pi)^{2-\beta} \int_0^\infty Q_{scat}(\chi, \mathfrak{M}) \chi^{1-\beta} \, d\chi \qquad (9.44)$$

Now, the integral is independent of wavelength, and has been shown to converge for value of $2 \le \beta \le 6$ (Van de Hulst, 1957), yielding $b_{scat} \propto \lambda^{2-\beta}$, which is a simple dependence indeed. Ångström (1929) and others observed the atmospheric extinction to be a power function of wavelength given by $b_{scat} \propto \lambda^{-\alpha}$, with $\alpha \simeq 1.2$ as a typical value.

Charlson (1972) performed measurements of *both* α and β in Los Angeles smog and showed that the simple relationship $\alpha = \beta - 2$ was nearly correct, and that it was adequate for many applications. However, small departures occurred due to the fact that the size distributions were not truly describable by a simple power law. The measurement system used—based on an integrating nephelometer—will be discussed later in this chapter.

9.7 Measurement on Aerosols

Most measurements of the composition and amount of atmospheric particulate matter have been accomplished with methods involving the removal of the particles from the air. There are numerous advantages to such methods. The acquisition of large samples makes it possible to use ordinary chemical techniques for analysis and analytical balances for determination of mass. The usual removal methods are inexpensive, and the use of ordinary laboratory appratus allows data acquisition without capital expenditure. The apparatus needed for removing particles from air (impingers, filters, etc. discussed in Chapter 3) are simple to operate. Perhaps one of the most frequent arguments for the continuance of these methods is that there is a large body of data from the past, and that identical methods (with reproducible errors) should be used if trends are to be detected

However, there are many disadvantages to this dependence on removal methods. The primary objection is that the particles are subjected to gross physical and chemical changes of their "micro environment," thus changing their composition and properties. Some particles may volatilize after removal, as suggested by Goetz *et al.* (1961), leading to errors in amount as well as composition. Other particles may react with the sampling surface or with other collected particles to form molecular species that did not exist in the atmosphere. For example, CaO or $CaCO_3$ dust could react with H_2SO_4 mist to neutralize the acid *and* the base even though the constituents might have come from different air parcels, sampled hours apart. Contamination of the sampled material by handling is possible. Ammonia gas in the laboratory could easily react with H_2SO_4 on a filter or impactor slide, altering the composition and mass.

Other types of problems exist in microscopic and electron microscopic analysis. The heating of the sample by the illumination in an optical microscope is sufficient to evaporate some particles. The optical microscope also fails to sense particles of inadequate size or scattering efficiency. Preparation for electron microscopy often involves coating or shadowing of the sample in a vacuum evaporation apparatus which tends to volatilize some substances. In electron microscopy or microprobe analysis, the electron beam tends to heat the particles and evaporate them. Even a relatively nonvolatile substance such as $(NH_4)_2SO_4$ has been shown to evaporate in an electron microscope (Heard and Wiffen, 1969).

One major problem exists in that most standard analytical procedures provide elemental or ionic information, not the molecular or crystalline composition. Since many of the important properties of particles are due to molecular rather than elemental character—for instance, toxicity, refractive index, density, particle shape, and physical state (liquid or solid)—more elaborate analytical techniques are needed. Simultaneously, more sophisticated sampling techniques and sample handling procedures (i.e., controlled atmosphere, cryogenic storage) are under study and may prove to be at least useful if not necessary for routine air pollution samples.

Present analytical methods in use in the National Air Sampling Network, which utilizes samples from high-volume filters, account for perhaps 25–30% of the weighable material. Since some of the sampled matter volatizes and is not even weighed, these data certainly indicate that changes in methodology can be anticipated, even though there may be merit in continuing some of the older methods for studies of trends (Wagman, 1970).

Given that (a) the removal of particles from air may change the character of the particles and (b) more sophisticated and elaborate approaches will be needed even for methods involving removal of particles from air, it seems advantageous to focus on those methods which sense physical and chemical properties of the particles *in situ*. Having dealt with the removal techniques in Chapter 3 and in the chapters on specific materials, the following discussion will focus on the sensing of aerosol properties *in situ*. Included are schemes where particles are removed from air and sensed immediately. Since these methods are all relatively new, changes and improvements can be anticipated in the near future.

9.7.1 ION MOBILITY

Historically, the study of particles in the atmosphere almost began with studies of electrical conductivity. The atmospheric particles were considered to fall into two size classes denoted small ions ($r < 10^{-3} \ \mu$m) and large ions, sometimes called Langevin ions ($10^{-3} < r < 10^{-1} \ \mu$m). Depending on the relative concentrations of these two classes of particles, their charge (sign and magnitude), and their respective mobility in air, the electrical conductivity is a measure of the content of sub-0.1 μm particulate matter. This natural property of air can be controlled and used to advantage for measuring size distribution.

Whitby *et al.* (1972a,b) have developed and refined a device for measuring particle sizes in atmospheric aerosol in the range $0.0075 \leq D_p \leq 0.6 \ \mu$m. Figure 9.17 shows the details of this device which has been used extensively in obtaining atmospheric data (see Figs. 9.6–9.8). The particles are first given a charge of controlled magnitude, and then are removed in the electric field of a mobility analyzer. The variation of charge passing through the analyzer with applied voltage on the analyzer is then interpreted in terms of size distribution.

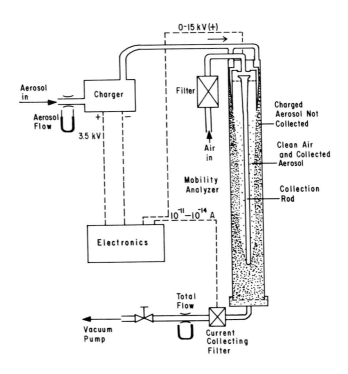

FIG. 9.17. The mobility analyzer and associated devices as developed by Whitby *et al.* (1972a,b).

This method is not truly *in situ* since particles are removed from air. Besides, some changes in size or composition of the particles might occur during the charging. These objections seem less important in this case, however, than in those removal techniques where properties are determined after removal, such as with filters for chemical analysis or impactors for microscopic or electron microscopic examination of the particles.

9.7.2 OPTICAL SINGLE PARTICLE COUNTERS

As shown in Fig. 9.15, the scattering efficiency factor is a strong function of size parameter. If this figure is replotted as the actual scattering cross section at a fixed wavelength as a function of size, a more or less monotonic increase of scattering with particle size results. The possibility exists that the same amount of light is scattered by particles of different size due to the oscillatory nature of the Q_{scat} versus χ curve.

In actual practice, the scattering at one angle is used rather than that for the range 0–180° as in Fig. 9.15. The same basic type of relationship exists, with the possibility of more than one particle size producing the same amount of light scattering. In all the various commercial versions of this device, the light scattered from each particle in a sample is sensed by a multiplier phototube as a pulse, the height of which is a function of particle size. A multichannel pulse-height analyzer segregates the pulses into channels and counts them, providing a size distribution. The practical range of such devices is $0.5 \lesssim D_p \lesssim 20$ μm, depending on configuration and operating characteristics. Other than heating of the air sample by the light source (which can be minimized), these types of devices are true *in situ* sensors that do not alter the nature of the particles. Commercial devices are available from Royco Instruments, Bausch and Lomb, and other companies. A discussion of the calibration of these devices has been published by Quenzel (1969a), while Whitby *et al.* (1972a,b) have demonstrated their application in atmospheric work.

9.7.3 ANGULAR INTEGRATING NEPHELOMETER

The term *integrating nephelometer* has come to imply a light scattering instrument that measures the sine-weighted angular integral [Eq. (9.36)]

$$b_{\text{scat}} = 2\pi \int_0^\infty \beta'(\phi) \sin \phi \, d\phi$$

which is the extinction component due to scattering by both particles and gas molecules. The first such instrument was devised by R.A.F. Lieutenant R. G. Beuttell in about 1943 as an aid to determination of visual range. Beuttell was killed in World War II, and his work was published posthumously by Prof. A. W. Brewer (Beuttell and Brewer, 1949).

Beuttell utilized the human eye as the sensor, and a matching technique was used to circumvent the lack of linearity in the response of the eye. Scattering coefficients down to $\sim 5 \times 10^{-4}$ m^{-1} could be detected. Detection implies a signal-to-noise ratio (S/N) of about unity. Ruppersberg (1967) devised an open device, again with a matching technique—a grey wedge—but with a pair of vacuum phototubes as the detectors and a xenon flash lamp as the light source. This combination improved the sensitivity by perhaps a factor of ten to a detection limit of below 1×10^{-4} m^{-1}. Crosby and Koerber (1963) utilized a single multiplier phototube and an optical unit in a ventilated van as well as a xenon flash lamp. This extended the detection limit to around 2×10^{-5} m^{-1}.

Ahlquist and Charlson (1967–1969), utilizing a multiplier phototube, solid-state circuitry, and an optical geometry similar to that of Crosby and Koerber, improved the signal-to-noise ratio by a factor of about 20 to yield a detection limit of less than 1×10^{-6} m^{-1}. This sensitivity, combined with a closed optical chamber of small volume (a few liters), permitted the use of filtered gases (air, CO_2, Freon 12, etc.) as standards of light scattering. The scattering coefficient of filtered air, for example, at a typical instrument wavelength of $\sim$500 nm is about 2×10^{-5} m^{-1}, and can be measured with a S/N of 20–100, depending on experimental conditions.

All of these devices were developed originally for the measurement of *meteorological range* L_v, which is defined as that distance where an ideal black object just disappears when viewed against the horizon sky in daytime. Assuming a homogeneous atmosphere, this quantity can be related to b_{scat} by the Koschmieder theory (Middleton, 1957).

$$L_v = 3.9/b_{scat} \qquad (9.45)$$

where the numerator ($3.9 = -\ln 0.02$) is derived empirically from the contrast threshold of the human eye. Of course, the wavelength chosen for measurement must match that of the sensitivity of the eye, around 500–550 nm.

The integrating nephelometer senses the scattering component of extinction b_{scat} by a particular geometry that provides the sine function in the integrand, without lenses or mirrors. Fig. 9.18a shows a simple nephelometer, indicating the key elements of the geometry. Figure 9.18b shows just the scattering geometry itself for definition of terms. The element of scattering volume dv is given by

$$dv = (r - x)^2 \omega \, dx \qquad (9.46)$$

The angles θ and ϕ are related by $\phi = \frac{1}{2}\pi - \theta$, where ϕ is the scattering

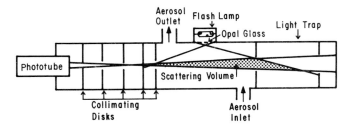

FIG. 9.18a.   Integrating nephelometer. The vertical dimension is exaggerated by a factor of about three in this drawing.

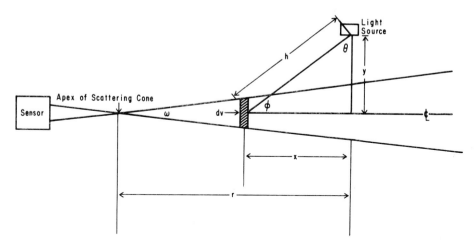

FIG. 9.18b. Geometry of the integrating nephelometer, exaggerated in y direction by $\sim 6:1$.

angle. Also,

$$x = y \cot \phi, \qquad dx = y \csc^2\phi \, d\phi$$

$$h = y \csc \phi, \qquad \cos \theta = \sin \phi$$

so

$$dv = (r - y \cot \phi)^2 \omega y \csc^2\phi \, d\phi$$

The illuminance at the scattering volume is given by (assuming that the opal glass has a cosine emission characteristic)

$$E = \frac{I_0 \cos \theta}{h^2} = \frac{I_0 \sin \phi}{y^2 \csc^2\phi} = \frac{I_0 \sin^3\phi}{y^2} \qquad (9.47)$$

The element of intensity from dv in the direction of the sensor is, from (9.35),

$$dI = E\beta'(\phi) \, dv \qquad (9.35a)$$

so

$$dI = [(I_0 \sin^3\phi)/y^2]\beta'(\phi)(r - y \cot\phi)^2 \omega y \csc^2\phi \, d\phi$$

$$= (I_0/y)(r - y \cot\phi)^2 \omega \beta'(\phi) \sin\phi \, d\phi \qquad (9.48)$$

But the luminance of the volume viewed from the apex of the viewing volume is inversely proportional to $\omega(r - x)^2$, so the element of luminance (intensity per unit area) $d\mathcal{B}$ is

$$d\mathcal{B} = (I_0/y) \int_{\phi_1}^{\phi_2} \beta'(\phi) \sin\phi \, d\phi \tag{9.49}$$

Thus, the sensor detects a luminance $\mathcal{B}$ given by

$$\mathcal{B} = (I_0/y) \int_{\phi_1}^{\phi_2} \beta'(\phi) \sin\phi \, d\phi \tag{9.50}$$

If the geometry is such that $\phi_1 \simeq 0$ and $\phi_2 \simeq \pi$, i.e., the device is very long, then

$$\mathcal{B} = (I_0/y) \int_0^{\pi} \beta'(\phi) \sin\phi \, d\phi = I_0 b_{\text{scat}}/2\pi y \tag{9.51}$$

The definitions of photometric quantities and this derivation follow from Middleton (1952), who concluded his analysis with the remark, "This happy result depends entirely on the cosine distribution of the radiation from the light source."

Thus we can see that Beuttell used two facts of geometric optics to obtain the particular integral he desired. First, the cosine characteristic of the source ultimately provides a $\sin\phi$ function in the integrand. Second, the conical viewing volume has a volume element that increases as $(r - x)^2$, and at the same time, the detector senses each volume element according to the inverse square law, or $(r - x)^{-2}$, which cancel out, yielding equal weighting of each linear increment dx.

One drawback to the use of all versions of the integrating nephelometer is the built-in angular truncation. Ideally, the device would collect light over an angular range from 0 to π (180°); unfortunately, the usual range from around 8° to about 170° introduces an error of perhaps 10% or less in a typical atmospheric case (Crosby and Koerber, 1963; Ensor and Waggoner, 1970; Quenzel, 1969b). The small magnitude of the error is surprising in view of the large amount of light scattered in the forward direction, that is, in the range from 0° to perhaps 20° or 30°. The essential factor in yielding such a small error is the existence of a sine function in the integrand, which has, of course, small magnitude near 0 and π (Middleton, 1952).

The simplest version of the integrating nephelometer, as shown in Fig. 9.18a, has its wavelength definition provided by a UV cutoff filter and the normal response of an S-11 photocathode.

Applications of this first device have been focused on three general areas—visibility, aerosol mass concentration, and background aerosol.

If the light scattering coefficients of pure gases are known, then they may be used to calibrate the device, yielding a capability for inferring meteorological range. Calibration is also possible via the use of scattering objects of known geometry (Middleton, 1952). Alternatively, the output of the device can be empirically correlated to observed visual range, and a calibration scale can then be based on the regression equation. Horvath and Noll (1969) studied the low-humidity case (less than $\sim 70\%$) and showed that the correlation coefficients are high (-0.9) between visual range and the light scattering coefficient for their location (Seattle). The regression equation agreed with the Koschmieder theory (after corrections regarding the effective wavelength) within about 10%. This suggests that the calibration with gases as performed for their experiments is adequately accurate for visibility purposes.

The application to mass concentration is based primarily on an empirical approach involving correlation of measured aerosol mass concentration ($\mu g/m^3$) with measured light scattering coefficient after subtracting Rayleigh scatter. Correlation coefficients for different urban locations range from 0.6 to 0.92, with an average value of 0.8 (Charlson et al., 1968). A theoretical approach to the relationship of mass concentration and scattering coefficient is possible, but is fraught with the necessity of many assumptions (Horvath and Charlson, 1969; Charlson, 1969). The theoretical approach does, however, demonstrate that the relationship of mass concentration and light scattering coefficient is, surprisingly, only weakly dependent on size distribution as long as an approximate power-law size distribution holds.

These two applications can be combined to demonstrate the relationship of visual range to aerosol mass concentration at low relative humidity. The product of visual range and mass concentration is $1.8^{+1.8}_{-0.9}$ g/m² for 90% confidence limits. While this same result has been obtained by more conventional visual range observations, the objectivity and automatic operation of the integrating nephelometer made possible the rapid acquisition of optical data with only a small amount of effort (Charlson, 1969).

Studies with this device in remote locations (e.g., Mt. Olympus, Washington, and Pt. Barrow, Alaska) show that there is a recognizable range of values of b_{scat} in such background situations. Porch et al. (1970) suggest on the basis of these measurements that a background level of aerosol exists, with $1.5 \leq b_{scat}/b_{Rayleigh} \leq 1.9$ representing a range for the geometric mean of the ratio at 500 nm. They further suggest a background aerosol mass concentration of from 1.6 to 16 $\mu g/m^3$ (90% confidence) or a geometric mean of from 4 to 8 $\mu g/m^3$ based on the above correlation of b_{scat} to mass concentration. Of course, it is necessary in such situations to subtract the Rayleigh scattering of the gas molecules prior to study of the aerosol contribution since this is large with respect to the scattering by aerosols.

The integrating nephelometer was easily adapted for fast ($\sim$1 sec) response by making three changes from the simpler version (Ahlquist and Charlson, 1968):

 1. A small instrument volume, ($\sim$2l) which allows faster change of the gaseous sample.

 2. The addition of a second phototube, which serves to compensate for time-dependent variations in flash-lamp intensity.

 3. The use of electronic analog division for compensation for flash-lamp brightness variation. This permits the use of the signal from *each* flash, rather than averaging over many flashes, thus allowing much faster response time.

Applications of this device include the rapid determination of spatial distribution of scattering coefficient. This in turn can be interpreted in terms of mass concentration. Ahlquist and Charlson (1968) show both horizontal profiles and vertical soundings to illustrate this capability. That the instrument was mounted on the *outside* of the single-engine aircraft (Cessna 180) is evidence for the lack of sensitivity to vibration.

Wavelength resolution was provided by the addition of a pair of filter wheels, each containing a set of four filters (one set for each of the two photosensors) and auxiliary circuitry (Ahlquist and Charlson, 1969). A typical set of filters includes 360 $\pm$ 10 nm, 436 $\pm$ 5 nm, 546 $\pm$ 5 nm, and 675 $\pm$ 20 nm, spanning the entire visible spectrum. The circuitry provides a continuous recording of the Ångström exponent α defined by

$$b_{\text{scat}} \propto \lambda^{-\alpha} \tag{9.52}$$

for wavelength λ, by use of the finite difference

$$\alpha = -d(\log b_{\text{scat}})/d(\log \lambda) \simeq -\Delta(\log b_{\text{scat}})/\Delta(\log \lambda) \tag{9.53}$$

Choice of a suitably ₋mall $\Delta \log \lambda$ permits an accuracy of $\pm$0.1 or less in α.

Data from this device have been accumulated in two urban locations and one unpolluted maritime site. The main result to date is that α is nearly constant for many types of aerosol, and varies only from 0.5 to 2 for virtually all cases studied. That α is often constant between different wavelength intervals also corroborates the utility of the Ångström formula. The Ångström exponent does *not* seem to exhibit short-term variations, even though b_{scat} does. Some of the implications of this observation related to the size distribution have already been discussed.

The most frequent value of α of 1.5–1.8 results in a brown color when a white object is viewed through haze. This can be demonstrated by viewing a white object such as a well-illuminated part of a cloud through a 1-mm thickness of some types of polyethylene which exhibit a similar dependence of light scattering on wavelength. The colors caused by light scattering with an α of 1.5–1.8 and that caused by NO_2 are similar (Charlson and Ahlquist, 1969; Waggoner and Charlson, 1971). Another consequence is that distant dark or black objects take on a blue cast which is *not* as wavelength-dependent as the blue sky (where α approaches four).

The regularity of α in urban air also has application for the use of the simpler wideband instrument. Experiments have shown that the monochromatic light scattering coefficient at all visible wavelengths is highly correlated (0.9–1.0) to that measured over a broad band from 420–550 nm. As a result, it is possible to infer the value of b_{scat} at any wavelength simply by use of the b_{scat}–λ dependence and the measured value of α.

In remote locations, such as the upper troposphere and stratosphere, it is possible for an aerosol to exist for some time with particles of nearly the same size. If such an aerosol is sufficiently dense and has just the right size distribution, bright objects such as the sun and moon may appear blue or

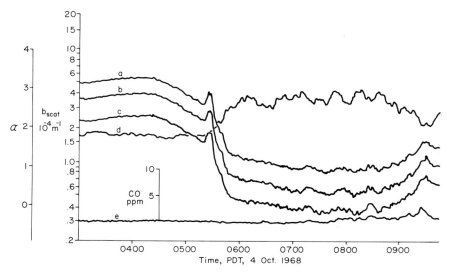

FIG. 9.19. Light scattering coefficient as a function of time for three wavelengths in Seattle, Washington. (Reprinted with permission from Ahlquist and Charlson, *Atmos. Environ.* **3,** 562 (1969), Pergamon Press.) (a) 436 nm; (b) 500 nm; (c) 600 nm; (d) wavelength exponent α; (e) CO concentration in ppm.

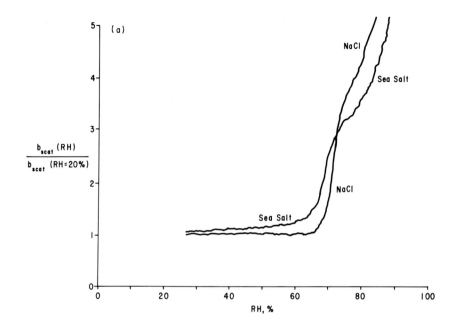

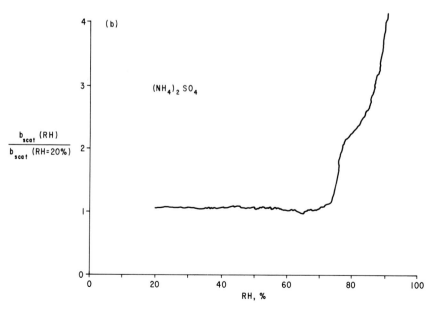

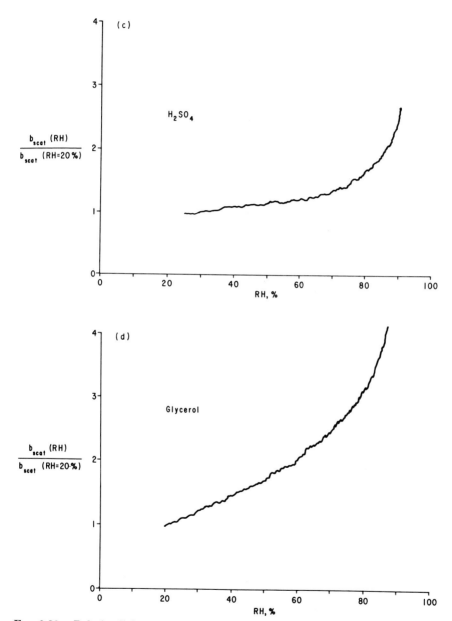

FIG. 9.20. Relative light scattering coefficient $b_{scat}(RH)/b_{scat}(RH = 20\%)$ versus relative humidity (RH) for laboratory aerosols generated in a bursting-bubble generator (Covert, 1972): (a) NaCl and sea salt aerosols; (b) $(NH)_2SO_4$; (c) H_2SO_4; (d) glycerol.

blue-green. Two spectacular occurrences have been documented, one following the gigantic volcanic eruption of Krakatoa in 1883, the other following large tundra fires in Canada in 1950 (Minnaert, 1954; Junge, 1963).

The possibility of continuously recording the aerosol component of b_{scat} and α alone (by subtracting the Rayleigh scattering) has shown that the condition of negative α is fairly frequent for the aerosol in very clean air. The rarity of the "blue moon" condition thus seems to be more nearly associated with the occurrence of high concentration of aerosol rather than with the occurrence of a particular size distribution. Since in the clean air situations, $b_{scat}/b_{Rayleigh} \simeq 1.5\text{--}2$ at sea level, the Rayleigh λ^{-4} dependence overrides a weakly negative α for the aerosol component alone (Porch et al., 1970).

Figure 9.19 shows a case of clean air replacing air of high aerosol content in Seattle, with the concomitant increase of α due to the increased importance of Rayleigh (λ^{-4}) scattering in the clean air.

The addition of a humidification step ahead of the integrating nephelometer allows the observation of the effects of relative humidity on light scattering. The humidifier can be a simple tube of 5 cm i.d., lined with wettable paper and wrapped with a heating tape. The paper can be kept wet by a small flow of water from a reservoir. The humidity may be measured by a psychrometer, the necessary flow velocity being provided by the sampling blower. A more complex automated system has been devised by Covert (1971) which utilizes a dew-point hygrometer modified electronically to yield relative humidity, an X–Y recorder, and an automatic humidity increasing system. Data on the b_{scat}–relative humidity dependence have been taken in urban air, in maritime air, and with pure laboratory aerosols. Figure 9.20 shows typical examples of laboratory results. Note in the curve for NaCl that the deliquescence begins at a humidity below that for bulk NaCl. This is due to the enhanced solubility of the smallest particles in the aerosol. The curve above $\sim 75\%$ RH is what would be expected from Raoult's equilibrium, as is the entire curve for H_2SO_4 and glycerol. The difference between sea salt and NaCl curves below 70% RH is probably due in large part to magnesium salts, the curve suggesting that sea salt particles are probably wet in most tropospheric conditions. Above 70%, the difference may be due to organic materials (Pueschel et al., 1969).

Figure 9.21 shows data on atmospheric aerosols, indicating two different responses. One aerosol is hygroscopic but not deliquescent, the other is hygrophobic. Figure 9.22 shows the response of b_{scat} of an H_2SO_4 aerosol before and after exposure to NH_3. The conversion to $(NH_4)_2SO_4$ would seem to have applications in identifying H_2SO_4 in air.

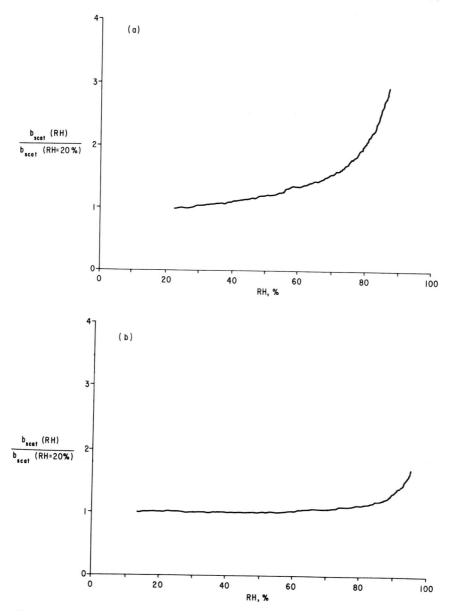

FIG. 9.21. Relative light scattering as a function of relative humidity as in Fig. 9.20 for two atmospheric situations. (a) Milpitas, California, 17 September 1971, 1245 PDT on a smoggy day. (b) Altadena, California, 23 September 1971, 1220 PDT in haze.

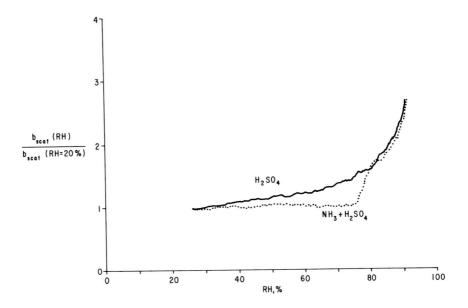

FIG. 9.22.   Relative light scattering as a function of relative humidity as in Fig. 9.20 for a laboratory H_2SO_4 aerosol before and after exposure to a low concentration of NH_3.

9.7.4   POLAR NEPHELOMETER

Devices which measure $\beta'(\phi)$ as a function of ϕ are called polar nephelometers. The integration of this quantity over angle provides the same information as the integrating nephelometer, though requiring more data handling. The major asset of the polar device—especially when polarization is also studied—is for simultaneous estimation of the size distribution and refractive index of the particles. This application requires knowledge or assumption of spherical particle shape, which is often a limitation.

Relatively few atmospheric studies have been made with this device, and so it is difficult to say if sufficient accuracy of refractive index determination is possible, for instance, for chemical analysis. Bullrich (1964) describes the apparatus and methods used in this method. It is also possible to measure $\beta'(\phi)$ for a single particle suspended in a Millikan apparatus (Blau et $al.$, 1970). It may be possible to obtain information regarding the refractive index and composition of the particles if sufficient accuracy can be achieved.

9.7.5 FLAME SCINTILLATION METHOD

When an aerosol particle enters a flame, it is heated, some or all of it evaporates, and the vapor may be raised to an excited electronic state. The subsequent emission of light is in some cases useful for both chemical identification and particle sizing. While this process certainly changes the composition of the particles, it does so in a short time period ($t < 10^{-3}$ sec), so that the history of the particles is fairly well known.

Pueschel (1969a,b) showed that some sodium-containing particles—specifically the halides—produce a flash of light when passing through a hydrogen flame the intensity of which is proportional to the square of particle radius. Other workers have shown that other flame conditions and compounds produce an intensity proportional to the cube of particle size. Control of the excitation conditions permits the use of this method for sensing specific elements in particles in some cases. Unfortunately, the determination of particle size depends on a knowledge of the composition of the particle. Since this is often unknown, the method is of limited utility in the atmosphere. However, for sea salt aerosol, which is over 90% NaCl, the method is definitely useful.

PROBLEMS

1. The power-law function, $dN/d \log r = Cr^{-\beta}$, has been used to describe aerosol size distributions. β is a parameter which varies from 2 to 4. Because of the difficulties with this function as r tends toward zero, a lower truncation limit must be defined. Evaluate the constant C, $\bar{r}$, and r_g, assuming a truncation limit of 0.01 μm, in terms of β and N_0 (the total number of particles/cm^3).

2. In a detailed investigation of the properties of an aerosol sample, the number concentration was 8.2×10^5 cm^{-3}, the mass concentration 9.9 μg/m^3, and the surface concentration 4.9×10^{-5} cm^2 cm^{-3}. Calculate r_g and $\ln \sigma_g$ for this sample assuming a log-normal size distribution and unit density.

3. Show that the Cunningham drag force expression reduces (a) to the molecular-kinetic one for large Kn and (b) to the Stokes formula for small Kn. State any assumptions that are necessary.

4. (a) Write the expression for settling velocity when Kn $\approx$ 1 and when Kn $\gg$ 1.

   (b) Develop a table of settling velocities for spherical particles of $\rho = 1$ at sea level in air.

5. Calculate the dust fall rate (mass time^{-1} area^{-1}) for the distribution in Fig. 9.4.
6. Show that the Köhler expression (9.1) reduces to the Kelvin formula (2.7) if there is no dissolved material.
7. Show that settling velocity is much smaller than the thermal velocity of particles under normal sea level conditions for particles of $r < 0.1$ μm.
8. Show that for large R and small t, the diffusion to a sphere reduces to that of a flat plane when $V_s = 0$.
9. Replot Fig. 9.15 as S versus r.
10. Show that σ_g of a log-normal size distribution is given by $r_{0.84}/r_g$ on a plot of the sort shown in Fig. 9.5b, where the logarithm of particle size is plotted against cumulative probability on a nonlinear scale.

REFERENCES

Ahlquist, N. C., and Charlson, R. J. (1967). *J. Air Pollut. Contr. Ass.* **17,** 467.
Ahlquist, N. C., and Charlson, R. J. (1968). *Environ. Sci. Technol.* **2,** 363.
Ahlquist, N. C., and Charlson, R. J. (1969). *Atmos. Environ.* **3,** 551.
Ångström, A. (1929). *Geograf. Ann.* **11,** 156.
Beuttell, R. G., and Brewer, A. W. (1949). *J. Sci. Instrum.* **26,** 357.
Blau, H. H., McCleese, D. J., and Watson, D. (1970). *Appl. Opt.* **9,** 2522.
Bolin, B. (1971). Sweden's case study contribution to the United Nations conference on the human environment. Inst. of Meterology, Univ. of Stockholm, Stockholm, Sweden.
Bullrich, K. (1964). Scattered radiation in the atmosphere" *Advan. Geophys.* **10,** 99.
Charlson, R. J. (1969). *Environ. Sci. Technol.* **3,** 913.
Charlson, R. J., and Ahlquist, N. C. (1969). *Atmos. Environ.* **3,** 653.
Charlson, R. J., Ahlquist, N. C., and Horvath, H. (1968). *Atmos. Environ.* **2,** 455.
Charlson, R. J. (and several other authors in a series of three papers) (1972). *J. Colloid Interface Sci.* **39,** 240.
Covert, D. S. (1971). A Study of the relationship of chemical composition and humidity to light scattering by aerosols, M. S. Thesis, Univ. of Washington, Seattle. [See also: Covert, D. S., Ahlquist, N. C., and Charlson, R. J. (1972). *J. Appl. Meteorol.,* **11.** To be published.]
Crosby, P., and Koerber, B. W. (1963). *J. Opt. Soc. Amer.* **53,** 358.
Davies, C. N., ed. (1966). "Aerosol Science." Academic Press, New York.
Einstein, A. (1908). *Z. Elektrochem.* **14,** 235.
Einstein, A. (1956). "Investigations on the Theory of the Brownian Movement." Dover, New York.
Ensor, D., and Waggoner, A. P. (1970). *Atmos. Environ.* **4,** 481.
Ensor, D. S., Charlson, R. J., Ahlquist, N. C., Whitby, K. T., Husar, R. B., and Liu, B. Y. H. (1972). *J. Colloid Interface Sci.* **39,** 242.

Fletcher, N. H. (1962). "The Physics of Rainclouds." Cambridge Univ. Press London and New York.

Fuchs, N. A. (1964). "The Mechanics of Aerosols." Pergamon, Oxford.

Goetz, A., Preining, O., and Kallai, T. (1961). *Geofis. Pura. Appl.* **50,** 67.

Heard, M. J., and Wiffen, R. D. (1969). *Atmos. Environ.* **3,** 337.

Herdan, G. (1960). "Small Particle Statistics," 2nd ed. Butterworth, London.

Hidy, G. M., and Brock, J. R. (1970a). "The Dynamics of Aerocolloidal Systems." Pergamon, Oxford.

Hidy, G. M., and Brock, J. R. (1970b). An assessment of the global sources of tropospheric aerosols. *Clean Air Congr. 2nd, Washington, D.C., December 1970.*

Horvath, H., and Charlson, R. J. (1969). *Amer. Ind. Hyg. Ass. J.* **30,** 500.

Horvath, H., and Noll, K. E. (1969). *Atmos. Environ.* **3,** 543.

Husar, R. B., Whitby, K. T., and Liu, B. Y. H. (1972). *J. Colloid Interface Sci.* **39,** 211.

Junge, C. E. (1963). "Air Chemistry and Radioactivity." Academic Press, New York.

Junge, C. E., and Abel, N. (1965). Modification of aerosol size distribution in the atmosphere and development of an ion counter of high sensitivity. Final Tech. Rep. No. DA 91-591-EUC-3483, DDCNo. AD469376. Johannes Gutenberg Univ., Mainz, Germany.

Kerker, M. (1969). "Scattering of Light and Other Electromagnetic Radiation." Academic Press, New York.

Köhler, H. (1921). *Geofys. Publ.* **2,** No. 3, 6.

Lamb, H. (1945). "Hydrodynamics," 6th ed. Dover, New York.

Lodge, Jr., J. P., Page, J. B., Basbergill, W., Swanson, G. S., Hill, K. C., Lorange, E., and Lazrus, A. L. (1968). Chemistry of United States precipitation Final. Rep. on the Nat. Precipitation Sampling Network. Nat. Center for Atmospheric Res., Boulder, Colorado.

Middleton, W. E. K. (1952). "Vision through the Atmosphere." Univ. of Toronto Press, Toronto, Canada.

Mie, G. (1908). *Ann. Phys. (Leipzig)* **25,** 377.

Minnaert, M. (1954). "The Nature of Light and Colour in the Open Air." Dover, New York.

Odén, S. (1969). "Regional Aspects of Environmental Disturbance" (in Swedish). *Vann 3,* Saertrykk No. 46. Johansen and Nielsen, Oslo, Norway.

Porch, W. M., Charlson, R J., and Radke, L. F. (1970). *Science* **170,** 315.

Pueschel, R. F. (1969a). Principles and applications of the flame scintillation spectral analysis, Ph.D. Dissertation, Univ. of Washington, Seattle.

Pueschel, R. F. (1969b). *J. Colloid Interface Sci.* **30,** 120.

Pueschel, R. F., Charlson, R. J., and Ahlquist, N. C. (1969). *J. Appl. Meteorol.* **8,** 995.

Quenzel, H. (1969a). *Appl. Opt.* **8,** 165.

Quenzel, H. (1969b). *Gerlands Beitr. Geophys.* **78,** 251.

Radke, L. F. (1970). *Proc. Second Int'l. Workshop on Condensation and Ice Nuclei,* L. O. Grant, ed. Colorado State Univ. Publication, Fort Collins, Colorado.

Robinson, E., and Robbins, R. C. (1971). Emissions concentrations and fate of particulate atmospheric pollutants. *Amer. Petrol. Inst. Publ.* **No. 4076.**

Rodhe, H., Persson, C., and Åkasson, O. (1971). Rep. AC-15, UDC 551.510.4. Inst. of Meteorology, Univ. of Stockholm, Stockholm, Sweden.

Ruppersberg, G. H. (1967). *Bull. Ass. Intern. Signalisation Maritime* **31,** 11 (in French).

Van de Hulst, H. C. (1957). "Light Scattering by Small Particles." Wiley, New York.

Waggoner, A. P., and Charlson, R. J. (1971). *Appl. Opt.* **10,** 957.

Wagman, J. (1970). Aerosol composition and component size distribution in urban atmospheres. *Conf. Methods in Air Pollution, 11th, 1970.* California State Dept. of Public Health, Berkeley.

Whitby, K. T., Husar, R. B., and Liu, B. Y. H. (1972a). *J. Colloid Interface Sci.* **39,** 177.

Whitby, K. T., Liu, B. Y. H., Husar, R. B., and Barsic, N. J. (1972b). *J. Colloid Interface Sci.* **39,** 136.

UNITS AND DIMENSIONS
USED IN AIR CHEMISTRY

The units used in this book to describe the composition of air may be divided into two basic categories: dimensionless and dimensioned. When there is a choice, dimensionless units are often preferred because of the freedom from problems in converting between one system of units and another (i.e., British and cgs or mks). However, in the case of some substances—notably particulate matter—dimensions are often a necessity. A common feature of dimensionless units is the use of certain multiples of ten times a fraction in describing small concentrations. Fraction, percent (%), parts per thousand (‰), parts per million (ppm), parts per hundred million (pphm), and parts per billion (ppb) are perhaps the most common units in use today. Normally, these dimensionless units imply a basis of molar, volume, or pressure measurement unless otherwise specified, e.g., "by weight." Since the number of molecules and the number of moles per volume in an ideal gas at a given condition is constant, the mole fraction may often be interpreted in terms of volume fraction, i.e., ppm (by volume) is the same as ppm by moles or by number of molecules. Similarly, the pressure fraction is the same as the mole fraction, etc.

It is perhaps unfortunate that the same nomenclature (ppm, ppb, etc.) is used in water chemistry to imply a dimensioned quantity of mass of solute or suspended solids per liter of water. One ppm is one milligram per liter, etc. As a result, those air chemists who often associate with aquatic chemists or who are concerned with being absolutely clear in their nomenclature append a "v" to the dimensionless symbols, ppmv, pphmv, ppbv, etc., to imply that volume is the basis of the unit.

Another common feature of the dimensionless units is that they are almost exclusively used for gaseous materials. While it would be possible to invent a dimensionless system for particulate matter (e.g. weight of particles per weight of air), it is not commonly done, and is not done in this book.

It is important to note the difference in the use of the term *billion*. In the United States and France, a billion is 10^9, while in the United Kingdom and Germany it is 10^{12}.

The list of units which follows is not exhaustive, but includes the quantities in common use today. The following symbols are used in the expressions for these units.

X_i —Mole fraction
V_i —Volume of ith component
P_i —Partial pressure of ith component
n_i —Number of moles of ith component
N_i —Number of molecules of ith component
m_i —Mass of ith component
P_i°—Saturation vapor pressure of ith component

1   Dimensionless Units for Gases

Basis	Unit	Symbol	Formula
Mole	Fraction	X_i	$n_i/\sum n_i$
	Percent	%	$100\,X_i$
	Parts per thousand	%o	$10^3\,X_i$
	Parts per million	ppm	$10^6\,X_i$
	Parts per hundred million	pphm	$10^8\,X_i$
	Parts per billion (U.S.)	ppb	$10^9\,X_i$
Volume*	Fraction, etc.	$\sim X_i$	$V_i/\sum V_i$
Pressure*	Fraction, etc.	$\sim X_i$	$P_i/\sum P_i$
Molecules*	Fraction, etc.	$\sim X_i$	$N_i/\sum N_i$
Weight†	Fraction	—	$m_i/\sum m_i$
	Parts per thousand	g/kg	$10^3\,m_i/\sum m_i$
Saturation ratio‡	Fraction	—	P_i/P_i°
	Percent	—	$100\,P_i/P_i^\circ$

* Dimensionless units for volume, pressure, or number of molecules are the same as for number of moles, since

$$X_i = V_i/\sum V_i = P_i/\sum P_i = N_i/\sum N_i \quad \text{if the gas is ideal}$$

† Called mixing ratio in meteorology.
‡ Called relative humidity, RH, for H_2O.

2 Dimensioned Units for Gases

Basis	Dimension	Units	Symbol
Partial pressure	Pressure, force (area)$^{-1}$	Torrecelli (mm Hg)	Torr
		Millibars	mb
		Atmospheres	atm
Mass concentration	Mass (volume)$^{-1}$	Micrograms per cubic meter	$\mu g/m^3$
Molarity	Moles (volume)$^{-1}$	Moles per liter	moles liter^{-1}
Molecular concentration	Molecules (volume)$^{-1}$	Molecules per cubic centimeter	cm^{-3}
Optical thickness	Length	Centimeters at standard T and P	cm

3 Dimensioned Units for Particulate Matter

Basis	Dimension	Units	Symbol
Mass concentration	Mass (volume)$^{-1}$	Micrograms per cubic meter	$\mu g/m^3$
Particle concentration	Number (volume)$^{-1}$	Particles per cubic centimeter	cm^{-3}
		Particles per liter	liter^{-1}
Surface area concentration	Area (volume)$^{-1}$	Square centimeters per cubic centimeter	cm^{-1}
Volume concentration	Volume (volume)$^{-1}$	Cubic micrometers per cubic centimeter	μm^3 cm^{-3}

4 Dimensioned Units for Particulate Matter Based on Effect

Basis	Units	Symbol
Condensation nuclei (CN) (active at RH $\sim$ 300%)	Number of CN per cubic centimeter	cm^{-3}
Cloud condensation nuclei (CCN) (active at RH $\leq$ 102%)	Number of CCN per cubic centimeter	cm^{-3}
Extinction of light	Extinction coefficient in exponential form of Beer-Lambert law	m^{-1}
Absorption of light	Absorption component of extinction coefficient	m^{-1}
Scattering of light	Scattering component of extinction coefficient	m^{-1}
Stain on filter paper	Coefficient of haze (an empirical quantity, not definable in metric units)	COH
Radioactivity	Picocuries per cubic meter	pcurie m^{-3}
Dustfall	Milligrams per square centimeter per month	mg cm^{-2} month^{-1}

LIST OF SYMBOLS

Symbol	Definition	First used in equation:
c	Concentration	1.1
t	Time	1.1
Q	Source term	1.1
R	Reaction terms	1.1
D	Diffusion term	1.1
P	Precipitation term	1.1
F	Flux density	1.2
D	Diffusion coefficient	1.2
F	Flux vector	1.3
∇	Del operator	1.3
∇^2	Laplacian operator	1.5
U	Velocity vector	1.6
$\bar{U}$	Average velocity	—
U'	Perturbation of velocity	—
$\bar{c}$	Average concentration	—
c'	Perturbation of concentration	—
$\bar{F}$	Average flux vector	1.7
K	Eddy dispersion coefficient	1.8
a, b	Coefficients in eddy dispersion equation	1.10
x, y	Horizontal direction coordinates	1.12
z	Vertical coordinate	1.12
K_x, K_y, K_z	Eddy dispersion coefficients in x, y, and z directions	1.12

Symbol	Definition	First used in equation:
σ_y	Standard deviation of plume in y direction	1.13
Q	Amount of material released at $t = 0$	1.10
σ_z	Standard deviation of plume in z direction	1.13
$\bar{U}$	Average velocity in x direction	1.13
Q'	Emission rate	1.13
h	Effective stack height	1.14
$\bar{c}_{box}$	Average concentration at downwind end of a box	1.16
h	Height of box or of mixed layer	1.16
P	Pressure	2.1
V	Volume	2.1
n	Number of moles	2.1
R	Gas constant	2.1
T	Absolute temperature	2.1
P_A	Partial pressure of A	—
n_A	Number of moles of A	—
P_{tot}	Total pressure	—
n_{tot}	Total number of moles	—
ρ	Density	2.2
M_A	Molecular weight of A	2.2
N_A	Number of molecules of A per cm^3	—
N_0	Avogadro's number	—
X_{lA}	Mole fraction of A in solution	2.3
$P_A{}^\circ$	Vapor pressure of A in its pure state	2.3
k_H	Henry's Law constant	2.4
$\bar{V}$	Volume per mole of condensed phase	2.5
σ	Surface tension	2.5
r	Radius of small particle	2.5
K_{eq}	Equilibrium constant	2.6
a_A, a_B, a_C	Activity of A, B, or C	2.6
G	Gibbs free energy	2.7
H	Enthalpy	2.8
S	Entropy	2.8
Δ	Forward difference operator	2.7
G°, H°, S°	Thermodynamic functions of state in the standard state	2.8
$G_{subscript}$, $H_{subscript}$, $S_{subscript}$	See Eq. (2.9) for explanation	—
subscript f	Formation	2.9
subscript T or a number	Temperature	2.9
I_λ	Intensity of light of wavelength λ	2.10
λ	Wavelength of light	2.10
z	Length of light path	2.10
α_λ	Molar extinction coefficient	2.10
I_0	Initial intensity	2.11
ϵ_λ	Absorptivity (decadic)	—
b_{abs}	Absorption component of extinction	2.12
b_{scat}	Scattering component of extinction	2.12
c_i	Concentration of ith substance	2.12

Symbol	Definition	*First used in equation:*
$\alpha_{a\lambda i}$	Molar absorption coefficient of ith substance at wavelength λ	2.12
$\alpha_{s\lambda i}$	Molar scattering coefficient of ith substance at wavelength λ	2.12
$(O_2), (NO_2), \ldots$	Concentration of $O_2, NO_2, \ldots$	—
k_f, k_r	Rate constant in forward and reverse directions	2.14
A	Preexponential factor	2.13
E_a	Activation energy	2.13
$h\nu$	Energy of a photon	—
ϕ	Primary quantum efficiency or quantum yield	2.15
I_a	Rate of photon absorption	2.15
F_0	Einsteins per area and time incident on the sample	2.16
Stk	Stokes's number	—
β	Figure of merit for filters	—
ϵ	Filter efficiency	—
V_0	Velocity in impactor	Fig. 3.11
L	Plate–orifice distance	Fig. 3.11
c_A'	Mass concentration of A	3.3
U	Flow rate	3.5
s	Standard error	4.1
x	A variable	4.1
$\bar{x}$	Average of x	4.1
n	Number of measurements of x	4.1
x_i	An individual measurement of x	4.1
$f(x)$	Fractional frequency of occurrence of a value of x per interval of x	4.2
σ	Standard deviation	4.2
σ_g	Geometric standard deviation	4.4
x_g	Geometric mean of x	4.4
$F(x)$	Fractional frequency of occurrence of a value of $\ln x$ per increment of $\ln x$	4.4
K_A	Partition ratio in chromatography	5.1
R_F	Retention time analog in thin-layer chromatography	5.2
P_a	Partial pressure of NH_3	6.1
K_{Ha}	Henry's law constant for NH_3	6.1
P_s	Partial pressure of SO_2	6.2
K_{Hs}	Henry's law constant for SO_2	6.2
P_c	Partial pressure of CO_2	6.3
K_{Hc}	Henry's law constant for CO_2	6.3
K_a	Equilibrium constant for NH_4OH dissociation	6.4
K_{1s}	Equilibrium constant for first H_2SO_3 dissociation	6.5
K_{2s}	Equilibrium constant for second (HSO_3^-) dissociation	6.6
K_{1c}	Equilibrium constant for first H_2CO_3 dissociation	6.7
K_{2c}	Equilibrium constant for second (HCO_3^-) dissociation	6.8
K_w	Equilibrium constant for water dissociation	6.9
a, b, c, d	Coefficients in pH calculation	6.11
M	Mass of atmosphere	—

Symbol	Definition	First used in equation:
ϕk_a	Photodissociation rate of NO_2	Ch. 7
k_3	Rate constant for $O + O_2 + M$ reaction	—
k_1	Rate constant for $O_3 + NO$ reaction	—
$\bar{V}$	Molar volume of liquid	9.1
σ	Surface tension	9.1
R	Gas constant	9.1
r	Drop size, particle size	9.1
i	Van't Hoff factor	9.1
m	Mass of solute	9.1
M_s	Molecular weight of solute	9.1
ρ	Density	9.1
$f(r)$	Fraction per unit of size r	—
$f(r)\,dr$	Fraction	—
$F(\ln r)$	Fraction per unit of $\ln r$	—
N	Total number of particles per volume	—
$f_m(r)$	Mass fraction per unit of r	9.2
$f_m(r)\,dr$	Mass fraction	9.2
$F_m(\ln r)$	Mass fraction per unit of $\ln r$	9.3
$f_v(r)$	Volume fraction per unit of r	—
$f_s(r)$	Surface fraction per unit of r	9.4
$f_l(r)$	Length fraction per unit of r	9.5
$f_q(r)$	qth moment fraction per unit of r	9.6
$\bar{x}$	Average of x	9.7
x_g	Geometric mean of x	9.8
σ_g	Geometric standard deviation	9.9
z, s	Dummy variables	—
r_{gq}	Geometric mean size evaluated from qth distribution	9.10
R_q	A specific integral	9.13
r^p	The pth power of r	9.16
β	Exponent of Junge distribution	9.18
D_p	Particle diameter	9.18
dN	Increment of number	9.18
l	Mean free path in gas	—
Kn	Knudsen number, l/r	—
F_{drag-1}	Stokes's drag force ($Kn \ll 1$)	9.19
η	Viscosity	9.19
V	Velocity	9.19
F_{drag-2}	Drag force when $Kn \simeq 1$	9.20
A	Cunningham slip correction	9.20
F_{drag-3}	Drag force when $Kn \gg 1$	9.21
m_g	Mass of gas molecule	9.21
k	Boltzmann constant	9.21
α_m	Empirical thermal accomodation coefficient	9.21
N_g	Number of gas molecules cm^{-3}	9.21
B	Mobility; velocity per unit force	9.22
g	Gravitational acceleration	9.23
V_s	Settling velocity	9.23

Symbol	Definition	First used in equation:
V_0	Initial velocity at $t = 0$	9.24
τ	Inertial period	9.24
$\mathcal{L}$	Stop distance	9.25
Stk	Stokes's number	9.26
l	Radius of curvature of streamlines	9.26
D	Fick diffusion coefficient	9.27
A	Surface area	9.28
N_{dep}	Number of particles deposited on a surface	9.28
N_0	Initial concentration	9.28
R	Radius of test sphere	9.29
K	Coagulation rate constant	9.33
I_ϕ	Scattered intensity at angle ϕ	9.34
ϕ	Scattering angle	9.34
E	Illuminance	9.34
R	Distance from scatterer	9.34
λ	Wavelength of light	9.34
$\mathfrak{M}$	Refractive index of scatterer	9.34
$\beta'(\phi)$	Volume scattering function	9.35
v	Volume of aerosol scatterer	9.35
b_{scat}	Scattering component of extinction (gas and particles)	9.36
b'_{scat}	Scattering component due to particles	9.37
S	Scattering cross section	9.37
b_{Rayliegh}	Rayleigh scattering component of extinction	—
N_i, S_i	Number and scattering cross section of ith particle class	9.38
Q_{scat}	Scattering area ratio	9.39
A	Particle area	9.39
χ	Size parameter, $2\pi r/\lambda$	—
$\mathfrak{M}_g$	Refractive index of a gas	9.40
f_{D}	Depolarization factor	9.40
b_{abs}	Absorption component of extinction	—
Q_{abs}	Absorption area ratio	9.41
b	Total extinction coefficient	9.41
C	Constant	9.43
α	Ångström exponent	—
L_{v}	Meteorological range	9.45
dv	Increment of scattering volume	9.46
r	Distance from light source to apex	9.46
ω	Solid angle	9.46
x	Distance from light source	9.46
θ	Complement of ϕ	9.47
h, y	See Fig. 9.18b	9.47
E	Illuminance	9.47
$\mathfrak{B}$	Luminance	9.49

GLOSSARY

Due to the fact that chemists, meteorologists, and other scientists are not usually familiar with each other's jargon, it is useful to collect a selected number of words which are used in many palces in this text and in the field of air chemistry. Most of these words are defined in one place in the text, and can be found in the index; however, this glossary is intended to serve as a quick reference for frequently used terminology. The reader is referred to textbooks in meteorology and chemistry for further elaboration. Of particular importance is "The Glossary of Meteorology" (Huschke, 1959).

REFERENCES

Huschke, R. E. (1959). "The Glossary of Meteorology." American Meteorological Society, Boston.

Middleton, W. E. K. (1952). "Vision through the Atmosphere." University of Toronto Press, Toronto.

Van de Hulst, H. C. (1957). "Light Scattering by Small Particles." Wiley, New York.

absorbance $Log_{10}I_0/I_\lambda$, where I_0 is the intensity incident upon an absorbing medium and I_λ the intensity transmitted through the medium.

absorption band   A *range* of wavelengths in which a substance absorbs light. Absorption bands are usually composed of a series of *absorption lines* which may be ascribed to individual energy transitions in the absorbing molecule.

219

absorption line   A narrow range of wavelengths in which a substance absorbs light.

accuracy   *See:* error, random; error, systematic; precision.

activity   The activity, a, and the activity coefficient, f, of a substance are defined by the equation $a = cf$ where c is the concentration (moles/liter). The activity coefficient, f, is usually an empirically determined quantity and depends not only on the concentration of the substance but on its particular properties and on the concentration and kind of other substances present. As solutions become ideal $(c \to 0)$, $f \to 1$.

adsorption, absorption   *See:* Sorption

advection   The transport of a quantity (of material, heat, or of other property) due solely to the motion of an air mass, i.e., transport due to the mean wind. Advection is usually considered to be a horizontal transport as opposed to convection, which involves local vertical motion, as in a cumulus cloud.

albedo   The ratio of the intensity of electromagnetic radiation reflected from a body (or surface) to the intensity incident upon it. Albedo connotes a broad wavelength band, while reflectivity more often is used for monochromatic radiation.

aliphatic compounds   Organic compounds which have an open-chain structure, as opposed to aromatic compounds. Examples of aliphatic hydrocarbons are butane, isopentane, octane, and other straight chain or branched chain molecules. *See also:* olefinic compounds; aromatic compounds.

amperometry   *See:* coulometry.

anthropogenic   Produced by human activities.

aromatic compounds   Organic compounds based on the ring structure of benzene

$$
\begin{array}{c}
\text{H} \\
\text{C} \\
\text{HC} \diagup \diagdown \text{CH} \\
\| \qquad | \\
\text{HC} \diagdown \diagup \text{CH} \\
\text{C} \\
\text{H}
\end{array}
$$

atmosphere   The envelope of air surrounding the earth and held to it by gravity. It is one of the three main elements of the environment; atmosphere, hydrosphere, and lithosphere. *See also:* hydrosphere; lithosphere.

attenuation   The diminution of a quantity. In electronics, the opposite of amplification. *See also:* extinction.

azo dye   A family of brightly colored compounds characterized by a linkage of aromatic groups by nitrogen atoms.

bimodal distribution   A plot of the frequency of occurrence of a variable versus the variable is a bimodal distribution if there are two maxima of the frequency of occurrence separated by a minimum. *See also:* mode.

biosphere   That spherical shell encompassing all forms of life on the earth. It extends from the ocean depths to a few thousand meters of altitude in the atmosphere, and includes the surface of land masses.

blue moon   A rare phenomenon which occurs when the particulate matter in the atmosphere is highly concentrated and scatters light more strongly in the red than in the blue.

brightness   *See:* luminance.

Brownian motion   The movement of particles in a colloidal system such as an aerosol caused by collision with the molecules in the fluid in which they are imbedded.

cascade impactor   An aerosol sampling device designed to separate particles into different size classes. The segregation depends on the fact that large particles have larger Stokes's numbers (Stk) than small ones, hence separation occurs due to differences in impaction characteristics. *See also:* Stokes's number.

chemiluminescence   The emission of light as a result of chemical reaction. An example of gas-phase chemiluminescence is

$$O_3 + NO \rightarrow NO_2{}^* + O_2$$
$$\downarrow$$
$$NO_2 + h\nu$$

chromatography   In general, the separation of substances by transporting them in a mobile phase over a stationary substrate for which they have differing affinities. The first syllable (chrom) was used originally in describing this method because the separations were usually detected as color differences in paper chromatography. Since no color difference is involved in gas chromatography, there is often a difficulty in relating the word to the method.

cloud   In general, a recognizable, visible aerosol parcel. Usually, *cloud* refers to a water droplet aerosol with relative humidity slightly above 100%, although dust clouds are also so identified.

coagulation   The process by which small particles in a colloidal system collide with and adhere to one another to form bigger particles. *Brownian motion* is the usual mechanism that causes collision, although others exist including sound waves and electrical forces.

colorimetry   A class of analytical methods which depends on the measurement of light absorption or depth of color in a solution.

conductimetry   Analytical methods that are based on the measurement of the electrical conductivity of a solution.

correlation coefficient (linear)   The linear correlation coefficient r, of two variables x and y, is defined by

$$r = \frac{\sum\limits_{i=1}^{n} (x_i - \bar{x})(y_i - \bar{y})}{\sqrt{\sum\limits_{i=1}^{n} (x_i - \bar{x})^2 \sum\limits_{i=1}^{n} (y_i - \bar{y})^2}}$$

where x_i and y_i are the measured values in the ith experiment of n total experiments. $\bar{x}$ and $\bar{y}$ are the arithmetic means of x_i and y_i

$$\left(\bar{x} = \frac{1}{n} \sum_{i=1}^{n} x_i \quad \text{etc.}\right)$$

In statistics, the linear correlation coefficient indicates the degree to which two quantities are linearly related. If $x = ay$, then $r = 1$, and departures from this relationship decrease r. The reader who is in need of using this statistical method is strongly urged to refer to a statistics text.

coulometry   An analytical method that is based on the measurement of the amount of electrical charge transferred in an electrochemical cell. Sometimes called *amperometry*.

cryogenic   Low temperature processes, apparatus, etc., are referred to as *cryogenic* when the temperatures are well below room conditions, usually at the temperature of liquid air or liquid N_2, or lower.

Cunningham slip correction   The correction factor applied to the Stokes drag force when the mean free path of the gas molecules is of the same scale as the particle size.

cyclic compound   An organic compound with a closed ring structure. Cyclopropane

$$\begin{array}{c} H_2 \\ C \\ H_2C \underline{\qquad} CH_2 \end{array}$$

is the simplest one. *See also:* polycyclic compounds: aromatic compounds.

dark reaction   A chemical reaction that proceeds in the absence of light.

deliquescence   The process that occurs when the vapor pressure of the saturated aqueous solution of a substance is less than the vapor pressure of water in the ambient air. Water vapor is collected until the substance is dissolved and in equilibrium with its environment.

deposition velocity   The ratio of the flux density ($g\ cm^{-2}\ sec^{-1}$) of a substance at a sink surface to the concentration in the atmosphere ($g\ cm^{-3}$). While the units of this ratio are clearly cm sec^{-1} and are thus a "velocity" it is important to realize that the ratio is not a flow velocity in the normal sense of the word.

diffraction   Forward scattering due to the wave nature of light. Diffraction is an edge effect which is especially important for larger particles in atmospheric aerosols, giving large scattered intensities at low scattering angles.

diffusion   A process by which substances, heat, or other properties of a medium are transferred from regions of higher concentration to regions of lower concentration. Molecular diffusion is caused by the Brownian motion of molecules, and results in a flux described by Fick's law [see Eq. (1.2)]. Eddy dispersion is often called eddy diffusion, and is due to the transport by fluid eddies.

diffusiophoresis   The motion of aerosol particles toward a surface where condensation is occurring or away from a surface where evaporation is in progress due to the forces of the diffusing species on the particles.

dry adiabatic lapse rate   The rate of temperature decrease with altitude in the absence of clouds for a parcel of air which is lofted adiabatically. $\Gamma = -\ dT/dZ = 9.8$ C/km for our terrestrial atmosphere.

dry fallout   *See:* fallout.

dustfall   *See:* fallout.

dustfall rate   The flux density of aerosol to a horizontal surface: dimensions (mass $area^{-1}\ time^{-1}$).

dynamic meteorology   The study of atmospheric motions, usually via mathematical methods, involving the equations of hydrodynamics. Sometimes called dynamics or fluid dynamics.

eddy   In turbulent fluid motion, a blob of the fluid that has some definitive character and moves in some way differently from the main flow.

eddy dispersion   The process by which substances are mixed in the atmosphere or in any fluid system due to the eddy motion. *See also:* diffusion.

efflorescence   The reverse process of deliquescence: the drying of a salt solution when the vapor pressure of the saturated solution of a substance is greater than that of the ambient air.

einstein   Avogadro's number of photons (6.023×10^{23} photons).

electrometry   A class of analytical methods employing electrical means for the quantitative measurement. *See also:* conductimetry, coulometry, potentiometry, voltametry.

eon 10^9 years.

error, random   The random fluctuations observed in the output from a measurement apparatus or method when the input to the instrument or method is held constant.

error, systematic   Errors in a measurement which stay more or less constant and which may be attributed to the particular design of the measurement. Such errors may in principle be accounted for and eliminated, in contrast to random errors, which may not be.

extinction   The attenuation of light due to scattering and absorption as it passes through a medium. *See:* Eq. (2.10): attenuation.

extinction coefficient   The quantity b in the Beer–Lambert law; $dI/I = - b\,dx$ where I is the intensity of light, dI is the attenuation over a path dx. Dimensions; $(\text{length})^{-1}$.

extraction   The removal of one substance (or class of substances) from one medium to another by means of a reagent in which the substance (or class of substances) is more soluble or in which a reactant is present which chemically binds the extracted material. In air chemistry, extraction means are often used to concentrate a trace gaseous constituent via chemical reactions in a scrubbing reagent.

fallout   The removal of particulate matter from air due to the effect of gravity. Also called sedimentation, dry fallout, or dustfall.

fluorometry   A group of analytical methods involving spectroscopic measurement of light emitted from a system which is excited by light. Ultraviolet light is usually used to excite the sample molecules to a state where emission occurs.

flux density   The flow of material or of a property of a medium across a unit of surface, in units of the quantity of material or of the property per area and time.

fog   Water droplet aerosol in contact with or close to the earth's surface. By international definition the aerosol is called a fog if it reduces visibility below 1 km. Fog differs from cloud only in its location.

fossil carbon, fossil fuels   Oil, coal, and associated gaseous materials sequestered in the earth's crust as fossil plants and other organisms.

fugacity   Fugacity is the analog for gaseous mixtures of activity in liquid solutions. In most cases, partial pressure is used to approximate fugacity.

Gaussian distribution function   A distribution function, $f(x)$, defined by the equation

$$f(x) = \frac{1}{\sigma\sqrt{2\pi}} \exp\left[-\frac{(x - \bar{x})^2}{2\sigma^2}\right]$$

where $\bar{x}$ is the arithmetic mean of x and σ is the standard deviation. *See:* Eq. 4.2.

grab sample A sample of air or of a substance extracted from air taken at a chosen point in time and space.

gravimetric methods A class of analytical methods, the basis of which is the determination of weight. A simple example is the precipitation of AgCl followed by drying and weighing in a Cl⁻ determination.

haze An atmospheric aerosol of sufficient concentration to be visible. The particles are so small that they cannot be seen individually, but are still effective in visual range restriction. *See:* visual range.

hydrocarbons Compounds containing only hydrogen and carbon. Examples; methane, benzene, decane, etc.

hydrometeor Any condensed water particle in the atmosphere of size much larger than the individual water molecule. Fog, cloud, some hazes, rain, snow, etc. are all hydrometeors.

hydrosphere The water portion of the earth (oceans, icecaps, lakes, rivers, etc.) as distinguished from the lithosphere and the atmosphere. *See:* lithosphere; atmosphere.

ideal gas A gas which obeys the ideal gas law; $PV = nRT$. Dry air comes very close to being an ideal gas.

illuminance Luminous energy crossing a unit surface area per unit time. Equivalent to luminous flux density. Dimensions; energy time^{-1} area^{-1}; Symbol: E. The illuminance is sometimes called intensity, as in Van de Hulst (1957).

impaction The process by which a particle is removed from a fluid with curved streamlines. The inertia of the particle causes it to not follow the fluid flow, but to move toward the fluid boundary.

imprecision *See:* error, random.

in situ An adjective from Latin meaning in its original position, in place.

intensity The energy flux per unit solid angle emanating from a source. Dimensions: energy (time)$^{-1}$ (solid angle)$^{-1}$; Symbol: I. (Not to be confused with illuminance, E. The relationship between I and E is

$$\frac{dI \cos \theta}{r^2} = dE$$

where r is the distance from the source and θ is the angle between incident rays and the normal to the receiving surface.) The use of intensity in the Beer–Lambert law is justified in the case of collimated light such as sunlight or such as in a spectrophotometer.

interception In aerosol physics, a mechanism whereby a particle collides with another or with an object by virtue of simple obstruction. A sieve intercepts particles larger than its hole size.

inversion, temperature In meteorology, a departure from the normal decrease of temperature with increasing altitude such that the temperature is higher at a given height in the inversion layer than would be expected from the temperature below the layer. This warmer layer leads to increased stability and limited vertical mixing of air. *See:* dry adiabatic lapse rate; stability.

isotropic A situation where a quantity (or its spatial derivatives) are independent of position or direction. Anisotropic and nonisotropic imply that the quantity is spatially dependent.

kinetics In chemistry, the study of the rates at which reactions occur and the influence of physical and chemical conditions on these rates.

lithometeor A particle of dry substance in the atmosphere, as contrasted to hydrometeor.

lithosphere The crust of the earth, usually thought of as discrete from and in contact with the hydrosphere and atmosphere. *See:* atmosphere; hydrosphere.

log-normal distribution A distribution function, $F(y)$, in which the logarithm of a quantity is normally distributed, i.e., $F(y) = f_{gauss}(\ln y)$ where $f_{gauss}(x)$ is a Gaussian distribution. *See:* Gaussian distribution: Eq. (9.9).

luminance "The luminance of a source in a given direction is equal to the illuminance produced on an elementary surface normal to this direction, divided by the solid angle subtended by the source at this surface. [Middleton (1952)]." Dimensions: energy (area)$^{-1}$ (solid angle)$^{-1}$.

mean (a) Arithmetic mean of x_i [*see* Eq. (9.7)]:

$$\bar{x} = \frac{\sum_{i=1}^{n} x_i}{n} \xrightarrow[n \to \infty]{} \int_0^\infty x f(x)\, dx$$

(b) Geometric mean of x_i [*see* Eq. (9.8)]:

$$x_g = \sqrt[n]{x_1 \cdot x_2 \cdot x_3 \ldots x_{n-1} \cdot x_n}$$

$$\ln x_g = \frac{\sum_{i=1}^{n} \ln x_i}{n} \xrightarrow[n \to \infty]{} \int_0^\infty \ln x\, f(x)\, dx$$

mean free path Mean free path describes a mean distance a particle travels in a given direction due to its thermal energy.

(a) For a molecule, mean free path l is given as $l = 1/\sqrt{2\pi} N d_m^2$ where N is the number of molecules per volume, d is their diameter, and l represents the mean distance between molecular collisons.

(b) For an aerosol particle, mean free path l_B in the Stokes region is $l_B = \bar{G}\tau = \bar{G}mB$ where $\bar{G}$ is the thermal velocity given by $\bar{G}^2 = 3kT/m$, m is the mass of the particle, k is Boltzmann's constant, T is the temperature and B is the mobility. *See:* mobility, Eq. (9.24).

mesoscale In meteorology, the size or scale of phenomena smaller than ordinary cyclones or weather systems but larger than such microscale phenomena as the thickness of the boundary layer, the wakes of objects, etc. Thunderstorms are mesoscale processes, and cities are usually mesoscale.

meteorological range The distance L_v given by Eq. (9.45);

$$L_v = \frac{3.9}{b_{scat}}$$

which is the distance over which an average observer could just see a large black object against the horizon sky during daytime, under isotropic conditions of b_{scat} and illumination. Dimensions: length.

micrometeorology The study of meteorological processes on scales from a millimeter or less up to tens or hundreds of meters.

Mie scattering   The scattering of electromagnetic radiation by spherical particles of any size r, relative to the wavelength, λ. Since the cases $r \ll \lambda$ and $r \gg \lambda$ are covered by Rayleigh (dipole) scattering and geometric scattering theories respectively, Mie scattering often refers to the case of $r \sim \lambda$.

mixing ratio   In meteorology, the dimensionless ratio of the mass of a substance (such as water vapor) in an air parcel to the mass of the remaining substances in the air. For trace substances, this is approximated by the ratio of the mass of the substance to the mass of air, but in the case of water vapor the mass of dry air is used. Units are sometimes used, such as g/kg.

mobility   In aerosol physics, the velocity of a particle per unit applied force. Dimensions; velocity/force; units: cm sec^{-1} dyne^{-1}.

mode   In a plot of the frequency of occurrence of a variable versus the variable, a maximum point is a mode.

molar absorptivity   The quantity ϵ_λ in the *decadic* form of the Beer–Lambert law:

$$I/I_0 = 10^{-\epsilon_\lambda c z}$$

where c is given in molar units, I is the transmitted intensity, and I_0 is the initial intensity. *See:* Eq. (2.11); molar extinction coefficient. *Note:* $\epsilon_\lambda = \alpha_\lambda/2.303$.

molar extinction coefficient   The quantity α_λ in the Beer–Lambert law: $I/I_0 = \exp(-\alpha_\lambda c z)$, where c is given in molar units, I is the transmitted intensity, and I_0 is the initial intensity. *See:* Eq. (2.11); molar absorptivity. *Note:* $\alpha_\lambda = 2.303\epsilon_\lambda$.

mole   Avogadro's number of atoms or molecules, N_0.

$$N_0 = 6.02252 \times 10^{23} \text{ molecules/mole}$$

nephelometry   Analytical methods which depend on the measurement of the intensity of scattered light emanating from an illuminated volume of a colloid (liquid or aerosol). The ratio of scattered intensity I to illuminating intensity I_0 is compared with a standard of known properties. *See:* turbidimetry.

normal distribution   *See:* Gaussian distribution function.

olefinic compounds   A special class of aliphatic hydrocarbons which contain one or more double bonds between carbon atoms. Examples are

Ethylene: $H_2C{=}CH_2$

2-Butene: $H_3CC(H){=}C(H){-}CH_3$

oxidized (oxidation)   In general, the state of an atom that has given up electrons in forming a molecule. Oxide formation is a specific case, as in the case of CO or CO_2, which are oxidized forms of carbon. *See:* reduced.

paraffin   A special class of aliphatic hydrocarbons containing only single carbon–carbon bonds. Examples are

Butane: $H_3C(CH_2)_2CH_3$

Isobutane: $H_3C{-}\underset{\underset{\displaystyle H}{|}}{\overset{\overset{\displaystyle CH_3}{|}}{C}}{-}CH_3$

phase In chemistry, a physically distinct, homogeneous portion of a heterogeneous mixture. There are three main phases—solid, liquid, and gas—the first two of which in many chemical systems have several separate phases such as those exhibited by ice or by immiscible liquids.

photochemistry The study of chemical reactions which result from exposure of a system to light.

photolysis Photochemical decomposition.

photometry Instrumental methods, including analytical methods, employing the measurement of light intensity.

photophoresis In aerosol physics, the motion of particles due to the influence of light. In many cases, this amounts to a special form of thermophoresis due to the heating of particles by the light.

photosynthesis The process by which plants convert CO_2 and H_2O to carbohydrates utilizing light as the energy source and producing O_2. The net reaction is:

$$nCO_2 + nH_2O + h\nu \rightarrow (CH_2O)_n + nO_2$$

pollutant Any atmospheric substance present in quantities greater than occur naturally, with attendant effects on man, materials, or the biosphere.

polycyclic compounds Cyclic compounds with more than one ring. Polycyclic aromatics are those substances with more than one benzene ring. Napthalene is a good example

potentiometry (a) In electronics, a potentiometer is a sensitive voltage measuring device using a null technique which provides infinite impedance at null.

(b) In chemistry, methods which involve the measurement with a potentiometer of voltage generated in a cell.

precipitation (a) In meteorology, rain or snowfall, etc.

(b) In chemistry, the sedimentation of a solid material from a liquid solution in which the material is present in amounts greater than permitted by its solubility.

precision The reproducability of output from an instrument or method with a constant input. *See:* random error.

rainout The removal of materials from air by rain. In rainout, incorporation of materials in hydrometeors within clouds implies a different mechanism from washout, which is a process occurring below cloud level.

Rayleigh scattering The scattering of light by particles much smaller than the wavelength. In the ideal case, the process is one of a pure dipole interaction with the electric field of the light wave.

reduced, reduction In general, the state of an atom or molecule that has taken up electrons. Reduced carbon compounds are things like coal, oil, and other hydrocarbons. Reduced sulfur compounds are molecules like methyl mercaptan (CH_3SH), hydrogen sulfide (H_2S), or elemental sulfur (S_8).

relative humidity The ratio of the partial pressure of water to the saturation vapor

pressure, also called saturation ratio

$$P_{H_2O}/P^\circ_{H_2O}$$

Relative humidity is often expressed as a percentage.

residence time   (a) Instrumental: The time required for air or reagent parcel to pass from the entrance to the exit of an instrument. Often approximated by the ratio of interior volume of the device to the flow rate.

   (b) Atmospheric: Residence time in the atmosphere is often estimated by the ratio of the average global concentration of a substance to its production rate.

saturation ratio   *See:* relative humidity.

scale   In meteorology, the size of the system. Microscale processes are the subject of micrometeorology, synoptic scale processes are studied in synoptic meteorology, etc.

scattering (light)   An interaction of a light wave with an object that causes the light to be redirected in its path. In elastic scattering, no energy is lost to the object.

scattering angle   The angle between the direction of propagation of the scattered and incident light (or transmitted light):

In this text ϕ is used for this angle and not θ as is common in physics texts. ϕ is used commonly in atmospheric optics. We consider only axially symmetric scattering.

scattering cross section   The extinction coefficient per particle, with units cm²/ particle. *See:* Eq. (9.37).

scavenging   The removal of materials from the gas or aerosol phase in the atmosphere into hydrometeors by cloud, rainout, or washout processes. Often called precipitation scavenging.

sedimentation   The removal of particulate matter from the atmosphere due to the effect of gravity. Also called *fallout, dry fallout,* or *dustfall.*

sink   In atmospheric chemistry, the sink is the receptor for material when it disappears or is removed from the atmosphere.

smog   Classically, a mixture of smoke plus fog. Today *smog* has a more general meaning of any anthropogenic haze. *Photochemical smog* involves the photolysis of NO_2 and other substances along with the visibility degrading haze.

solubility   The amount of a material that will dissolve in a given amount of a solvent, often in grams/100 ml.

solution   A mixture in which the components are uniformly distributed on an atomic or molecular scale. While liquid, solid and gaseous solutions exist, common nomenclature implies the liquid phase unless otherwise specified.

sorption   A class of processes by which one material is taken up by another. Absorption implies the penetration of one material into another; adsorption involves a surface phenomenon.

source   In atmospheric chemistry, the place, places, group of sites, or areas where a substance is injected into the atmosphere. Point sources, elevated sources, area sources, multiple sources, etc., are so identified.

spectral   An adjective implying a separation of wavelengths of light or other waves into a spectrum or separated series of wavelengths.

spectrometry   Analytical methods involving the measurement of wavelength characteristics of light, yielding information about the source of the light or the medium through which it has passed.

spectrophotometry   Methods using a measurement of light intensity with a photomoter as a function of wavelength.

static stability   In meteorology, the stability of the atmosphere in the vertical direction, to vertical displacements. If Γ is the dry adiabatic lapse rate, dry air is stable if the lapse rate $[-dT/dZ] < \Gamma$, unstable if $[-dT/dZ] > \Gamma$, and neutral if $[-dT/dZ] = \Gamma$. *See:* dry adiabatic lapse rate; inversion.

standard deviation   *See:* Gaussian distribution function.

stoichiometric proportions   When two or more substances are allowed to react, and the proportions are such that there is no excess of any reactant, the proportions used are stoichiometric. Stoichiometry is the study of proportions of substances in reacting systems.

Stokes's number   Sometimes referred to as the inertial parameter, Stk is an index of the impactability of an aerosol particle. *See:* Eq. (9.26).

stratosphere   The atmospheric shell lying just above the *troposphere*, characterized by a stable lapse rate. The temperature is approximately constant in the lower part of the stratosphere, and increases to a height of $\sim$50 km, which defines the top of the stratosphere and the bottom of the mesosphere.

Suess effect   The dilution of natural $^{14}CO_2$ by $^{12}CO_2$ due to human consumption of fossil fuels which contain essentially no ^{14}C. *See:* Chapter 8.

supersaturation   (a) In chemistry, a system is supersaturated with respect to a substance if it is present in a concentration greater than that permitted by its solubility (in a liquid) or by the vapor pressure (in a gas) of its condensed phase.

   (b) In meteorology, supersaturation with respect to H_2O is quantitatively given by the saturation ratio minus one, or the percent supersaturation is the percent relative humidity minus 100.

synoptic scale   In meteorology, the size or scale of ordinary weather systems or cyclones; typically 1000 km.

teragram, Tg 10^{12} grams.

terpenes   A group of hydrocarbons produced by trees and plants and subsequently emitted into the atmosphere. Examples: α-pinene and limonene. *See:* Chapter 8.

thermal force   In aerosol physics, the force on a particle in a gas due to a temperature gradient in the gas. In general, the particles move from warm to cool regions in the gas, and may be deposited on a cold surface in contact with warmer aerosol. Different theories exist depending on the relationship of particle size to mean free path in the gas phase.

thermal precipitator   An aerosol sampling device utilizing the thermal force to remove particles from the gas to a sampling surface. Temperature gradients of 100 C/mm or greater are often used, with rather low flow rates.

thermophoresis   The motion of aerosol particles in a thermal field.

titration   The addition of a standard solution (or standard gas mixture) to a fixed volume of unknown solution (or gas) usually from a buret, until the amount added is chemically equivalent to the substance being measured. This point is the end point, and is detected in many cases with an indicator that senses a miniscule excess of the standard solution.

titrimetric methods A class of volumetric analytical methods employing titration as the basis of quantitative measurement.

transmittance (transmission) The ratio of transmitted intensity to initial intensity, usually at a given wavelength, I_λ/I_0.

troposphere The lowest layer of the atmosphere, ranging from the ground to the base of the stratosphere at 10–15 km of altitude depending on latitude and weather conditions. About 70% of the mass of the atmosphere is in the troposphere where most weather features occur.

turbidimetry Analytical methods for colloidal systems where the transmittance of the turbid medium is measured. *See:* nephelometry.

van't Hoff factor A factor in solution chemistry that is used to account for the dissociation of salts and electrolytes and for the nonideality of concentrated solutions. In the case of dilute solutions, the van't Hoff factor i approaches a whole number given by the number of ions produced by the dissociation of a salt or electrolyte. For NaCl, $i \rightarrow 2$; for H_2SO_4, $i \rightarrow 3$; etc.

vapor A gaseous substance below its critical temperature which may be liquefied by pressure alone. Vapors may also be liquefied by cooling. It is important to note that vapors are gases, and that they are generally not visible in air. Common usage of the term water vapor to describe haze is clearly incorrect.

visual range The distance at which a large black object just disappears from view. Also called visibility. *See:* meteorological range.

volumetric methods A group of analytical methods in which the basis of measurement is a volume determination, such as in *titration*.

washout The removal of material from air by precipitation scavenging below the cloud base, for instance by falling raindrops.

AUTHOR INDEX

Numbers in italics refer to the pages on which the comple references are listed.

A

Abel, N., 183, 196, *209*
Adams, D. F., 43, *60*, 112, *113*
Adlard, E. R., 79, *98*
Ahlquist, N. C., 39, *60*, 189, 190, 199, 200, 201,204, *208*
Åkasson, O., 164, *209*
Albrecht, B., 71, *75*, 116, *132*
Allen, E. R., 105, *114*
Altshuller, A. P., 41, *60*, 117, 120, *131*, 147, 154, *155*
Ammons, B. E., 53, *61*
Ångström, A., 191, *208*
Arnold, P. W., 116, 122, *131*
Axelrod, H. D., 45, *60*

B

Baker, C., 90, *98*
Bamesberger, W. L., 112, *113*
Barnes, H. M., 91, *98*
Barrett, E. W., 39, *60*, 96, *98*
Barsic, N. J., 171, 172, 174, 193, 194, 195, *210*
Basbergill, W., 164, *209*
Bates, D. R., 116, *131*
Beck, R., 71, *75*, 116, *132*
Begeman, C. R., 140, *155*
Bellamy, L. J., *97*
Bellar, T. A., 147, *155*
Belsky, T., 154, *155*

Ben-Dov, O., 39, *60*, 96, *98*
Benson, S. W., *37*
Berkner, L. V., 6, *23*
Beuttell, R. G., 195, *208*
Biller, W. F., 42, *61*
Bischof, W., 138, *155*
Bischoff, K. B., 66, *75*
Blau, H. H., 206, *208*
Bock, R., 123, 124, *131*
Bodenstein, M., 130, *131*
Bolin, B., 137, 138, *155*, 164, *208*
Borg, K. M., 153, *155*
Brame, E. W., 92, *98*
Bratzel, M. P., 87, 88, *99*
Braverman, M. M., 111, *113*
Brewer, A. W., 195, *208*
Brock, J. R., 157, 158, 159, 162, 177, *209*
Brody, S. S., 112, *113*
Broecker, W. S., 6, *23*
Brown, R., 95, *98*
Bufalini, J. J., 117, 120, *131*
Bullrich, K., 186, 187, 206, *208*
Burg, W. R., 59, *61*, 124, *132*
Businger, J. A., 13, *23*, 141, *155*
Butcher, S. S., 40, *60*

C

Cadle, R. D., 105, *114*
Callendar, G. S., 136, *155*
Calvert, J. G., 35, *37*
Chaney, J. E., 112, *113*

231

Charlson, R. J., 10, *23*, 39, 40, *60*, *61*, 71,
    73, *75*, 189, 190, 191, 196, 199, 200,
    201, 204, *208*, *209*, *210*
Cheek, C. H., 5, *24*, 143, *155*
Christian, C. M., 87, *98*
Christian, G. D., *97*
Clingenpeel, J. M., 125, *132*
Clyne, M. A. A., 117, *131*
Cohen, I. R., 41, *60*
Cohen, N., 120, *132*, 142, *155*
Colucci, J. M., 140, *155*
Conley, R. T., *98*
Coomber, J. W., 120, *131*
Covert, D. S., 204, *208*
Craig, R. A., 122, *131*
Craw, A. R., 38, *60*
Crider, W. L., 111, *113*
Crosby, H. J., 36, *37*
Crosby, P., 195, 198, *208*

D

Dal Nogare, S., *98*
Darley, E. F., 79, *98*
Davidson, N., 130, *132*
Davies, C. N., 52, *60*, 158, *208*
Davies, J. H., 96, *98*
Delwiche, C. C., 115, *131*
Dickerson, R. C., 38, *61*
Dietz, V. R., 44, *60*
Doerr, R. C., 90, *99*
Dubois, L., 86, 90, *98*, *99*
Dworetzky, L. H., 140, *155*

E

Einstein, A., 181, 182, *208*
Eisenhart, C., 64, *75*
Elbert, W. C., 81, *99*
Elfers, L. A., 64, *75*, 109, 112, *114*
Elskin, R. H., 92, *98*
Engelmann, R. J., 22, *23*
Ensor, D. S., 189, 190, 198, *208*
Eriksson, E., 103, *113*
Ettre, L. S., 79, 95, *98*
Evans, W. H., 31, *37*

F

Feldman, F. J., *97*
Feldmann, C. R., 59, *61*
Fiegl, F., 81, *98*
Fielder, R. S., 110, *113*
Flath, R. A., 92, *98*
Fleagle, R. G., 13, *23*, 141, *155*
Fletcher, N. H., 22, *23*, 159, *209*
Fontijn, A., 128, 129, *131*
Forler, S. H., 71, 73, *75*
Frank, E. R., 50, *61*
Friedlander, S. K., 120, *131*, 148, *155*
Friend, J. P., 10, *23*, 105, *113*
Fritz, J. S., 110, *113*
Frost, A. A., *37*
Fuchs, N. A., 51, 52, *61*, 158, 177, 179,
    181, *209*

G

Gaeke, G. C., 108, *114*
Gelman, C., 50, *61*
Georgii, H. W., 104, 105, *113*, 123, *131*
Gerhard, E. R., 105, *113*
Gilbert, N., 125, *132*
Glasson, W. A., 130, *131*
Goetz, A., 192, *209*
Gohlke, R. S., 95, *98*
Goldsmith, J. R., 142, *155*
Graul, R. J., 154, *155*
Greenberg, L., 109, *113*
Guinn, V. P., 93, *98*

H

Hailey, D. M., 91, *98*
Hanst, P. L., 90, *99*
Harrison, G. R., *98*
Harrison, H., 122, *131*
Hauser, T. R., 90, *98*
Hays, D. B., 116, *131*
Healy, T. V., 5, *23*, 106, *113*, 116, *131*
Heard, M. J., 192, *209*
Heftmann, E., *98*
Heller, A. N., 140, *155*
Helweg, H. L., 154, *155*
Henderson, N., 92, *98*
Herdan, G., 171, *209*
Herzberg, G., 83, *98*
Hexter, A. C., 142, *155*

Hidy, G. M., 148, *155*, 157, 158, 159, 162, 177, *209*
Hill, K. C., 164, *209*
Himmelblau, D. M., 66, *75*
Hinkley, E. D., 96, *98*
Hobbs, P. V., 106, *114*
Hochheiser, S., 109, 111, *113*, 124, *131*
Hodgeson, J. A., 128, *131*
Horvath, H., 199, *208*, *209*
Hoshino, H., 43, *61*
Howard, R., 42, *61*, 95, *99*
Hughes, E., 5, 6, *24*
Hurn, R. W., 125, *132*
Husar, R. B., 171, 172, 174, 175, 184, 189, 190, 193, 194, 195, *209*, *210*

I

Imada, M., 154, *155*
Inaba, H., 96, *98*

J

Jacobs, M. B., 109, 111, *113*, 124, *131*
Jaffe, I., 31, *37*
Jeffries, H. E., 143, *155*
Johnson, K. L., 140, *155*
Johnson, W. B., 74, *75*, 96, *98*
Johnston, H., 122, *131*
Johnston, H. S., 36, *37*
Johnstone, H. F., 105, *113*
Jolly, W. L., 42, *61*
Junge, C. E., 71, 72, 73, *75*, 101, 102, 104, 105, *113*, *114*, 116, *132*, 139, 141, *155*, 164, 174, 183, 204, *209*
Jungroth, D. M., 43, *60*
Juvet, R. S., *98*

K

Kallai, T., 192, *209*
Keeling, C. D., 136, *155*
Kelley, P. L., 96, *98*
Kellogg, W. W., 105, *114*
Kerker, M., 185, *209*
Kettner, K. A., 79, *98*
Keulemans, A. I. M., *98*
Klotz, I. M., 29, *37*
Kobayasi, T., 96, *98*
Köhler, H., 159, *209*
Koerber, B. W., 195, 198, *208*

Kopczynski, S. L., 41, *60*
Koppe, R. K., 43, *60*
Kothny, E. L., 154, *155*
Koyama, T., 143, *155*
Kramer, G. D., 38, *61*
Krost, K. J., 128, *131*
Kummer, W. A., 129, *131*

L

LaHue, M. D., 5, *24*, 116, 123, *131*
Lamb, H., 177, *209*
Larsen, R. I., 72, *75*
Lazrus, A. L., 105, *114*, 164, *209*
Leavitt, J. M., 38, *61*
Lederer, E., *98*
Lederer, M., *98*
Leighton, P. A., 35, *37*, 117, 120, *131*
Levine, S., 31, *37*
Lewis, J. B., 43, *61*
Linnenbom, V. J., 5, *24*, 143, *155*
Littlewood, A. B., *98*
Liu, B. Y. H., 171, 172, 174, 175, 184, 189, 190, 193, 194, 195, *209*, *210*
Lodge, J. P., 45, 50, 53, *60*, *61*, 116, 123, *131*
Lodge, J. P., Jr., 5, *24*, 164, *209*
Loftin, H. P., 87, *98*
Lonneman, W. A., 41, *60*, 147, *155*
Loofbourow, J. R., *98*
Lorange, E., 164, *209*
Lord, R. C., *98*
Lovelock, J. E., 58, *61*, 79, *98*
Lowry, T., 119, *132*
Lukens, H. R., 93, *98*
Lundin, R. E., 92, *98*
Lutrick, D., 73, *75*

M

McCleese, D. J., 206, *208*
MacCready, P. B., 39, *60*
McFadden, W. H., 79, 95, *98*
Machta, L., 5, 6, *24*
McKay, H. A. C., 5, *23*, 106, *113*, 116, *131*
McKee, H. C., 44, *61*
Maggs, R. J., 79, *98*
Manganelli, R. M., 109, *114*
Mansfield, J. M., 87, 88, *99*
Manson, J. E., 104, *113*

Mantell, C. L., 44, *61*
Marshall, L. C., 6, *23*
Martell, E. A., 105, *114*
Mason, B. J., 21, *24*, 105, *114*
Metro, S. J., 43, *61*
Middleton, W. E. K., 187, 196, 198, 199, *209*
Mie, G., 185, *209*
Millan, M. M., 39, *61*, 96, 97, *98*
Miller, D. L., 154, *155*
Miller, M. S., 148, *155*
Minnaert, M., 204, *209*
Mitteldorf, A. J., 88, *98*
Moffat, A. J., 39, *61*, 97, *98*
Mon, T. R., 92, *98*
Monkman, J. L., 86, 90, *98*, *99*
Moore, G. E., 86, *99*
Moore, H., 154, *155*
Moore, W. J., *37*
Morgan, C. H., 110, *113*
Mueller, P. K., 149, 154, *155*

N

Nauman, R. V., 108, *114*
Nederbragt, G. W., 129, *132*
Nelson, C. J., 64, *75*, 112, *114*
Newcomb, G. S., 96, 97, *98*
Noll, K. E., 199, *209*
Norris, C. H., 64, *75*, 112, *114*
Norris, D., 109, *114*

O

Odén, S., 164, *209*
O'Keeffe, A. E., 58, 59, *61*, 112, *114*, 128, *131*
Ortman, G. C., 58, *61*, 112, *114*

P

Page, J. B., 164, *209*
Pales, J. C., 136, *155*
Palmer, H. F., 64, *75*, 112, *114*
Pate, J. B., 5, *24*, 53, *61*, 116, 123, *131*
Pearson, R. G., *37*
Persson, C., 164, *209*
Pfaff, J. D., 81, *99*
Pich, J., 47, *61*
Pierce, L. B., 154, *155*
Pilbeam, A., 5, *23*, 106, *113*, 116, *131*

Pitts, J. N., 35, *37*, 120, 129, *131*
Pooler, F., 38, *61*
Porch, W. M., 199, 204, *209*
Preining, O., 192, *209*
Pressman, J., 142, *155*
Prophet, H., 31, *37*
Pueschel, R. F., 204, 207, *209*

Q

Quenzel, H., 195, 198, *209*
Quiram, E. R., 42, 43, *61*

R

Radke, L. F., 161, 199, 204, *209*
Ramaswamy, G., 59, *61*, 124, *132*
Randerath, K., *98*
Rasmussen, R. A., 41, *61*, 143, 144, *155*
Regener, V. H., 122, 128, *132*
Rehme, K. A., 63, *75*
Revelle, R., 137, *155*
Ripley, D. L., 125, *132*
Ripperton, L. A., 143, *155*
Robbins, R. C., 101, 105, *114*, 115, 117, 122, *132*, 143, 153, *155*, 162, *209*
Roberts, L. R., 44, *61*
Robertson, T. J., 112, *113*
Robinson, E., 101, 105, *114*, 115, 117, 122, *132*, 143, 153, *155*, 162, *209*
Robinson, J. W., 87, 91, *98*
Rock, S. M., 42, *61*, 95, *99*
Rodes, C. E., 64, *75*, 112, *114*
Rodhe, H., 164, *209*
Ronco, R. J., 128, 129, *131*
Rosenberg, E., 63, *75*
Rossini, F. D., 31, *37*
Rubey, W. W., 3, *24*
Ruff, R. E., 40, *60*
Ruppersberg, G. H., 195, *209*
Ryan, T. G., 105, *113*

S

Sabadell, A. J., 128, 129, *131*
Saltzman, B. E., 59, *61*, 63, *75*, 124, 125, 126, *132*
Sawicki, E., 81, 90, *98*, *99*
Scargill, D., 5, *23*, 106, *113*, 116, *131*

Scaringelli, F. P., 63, *75*, 109, *114*
Schott, G., 130, *132*
Schuck, E. A., 117, 121, *132*
Schütz, K., 71, *75*, 116, 123, 124, *131*, *132*
Schuman, L. M., 119, *132*
Scott, W. D., 105, 106, *114*
Scott, W. E., 90, *99*
Seiler, W., 139, 141, *155*
Seinfeld, J. H., 120, *131*
Selvidge, H., 39, *60*
Sheesley, D. C., 50, *61*
Shepherd, M., 42, *61*, 95, *99*
Sinke, G. C., 31, *37*
Slade, D. H., 13, 16, 18, 19, *24*
Sleva, S. F., 41, *60*, 154, *155*
Slinn, W. G. N., 22, *23*
Smith, W. J., 48, *61*
Spurney, K. R., 50, *61*
Stahl, E., *98*
Stalker, W. W., 38, *61*
Stanley, T. W., 81, 90, *98*, *99*
Steer, R. P., 129, *131*
Stephens, E. R., 79, 90, *98*, *99*, 117, 121, *132*
Stevens, R. K., 112, *114*, 128, *131*
Stormes, J., 42, *61*, 95, *99*
Stull, D. R., 31, *37*
Suess, H. E., 137, *155*
Surprenant, N. F., 48, *61*
Sutterfield, F. D., 41, *60*
Swanson, G. A., 53, *61*
Swanson, G. S., 164, *209*
Swinnerton, J. W., 5, *24*, 143, *155*

T

Teck, R. J., 45, *60*
Teranishi, R., 92, *98*
Terraglio, F. P., 109, *114*
Thomas, R. S., 86, *99*
Thrush, B. A., 117, *131*
Tron, F., 108, *114*
Tsuchiya, T., 43, *61*
Tuesday, C. S., 130, *131*
Tukey, J. W., 137, *155*
Turner, D. B., 20, *24*

U

Udenfriend, S., *98*
Uthe, E. E., 74, *75*, 96, *98*

V

Van de Hulst, H. C., 185, 191, *209*
Van den Heuvel, A. P., 105, *114*
van der Horst, A., 129, *132*
van Diujn, J., 129, *132*
Van Valen, L., 6, *24*
Vossen, P. G. T., 95, *98*

W

Waggoner, A. P., 198, 201, *208*, *210*
Wagman, D. D., 31, *37*
Wagman, J., 193, *210*
Wanta, R. C., 38, *61*
Warneck, P., 139, 141, 142, 149, *155*
Wartburg, A. F., 41, 45, *60*, 126, *132*
Wasada, N., 43, *61*
Washburn, E. W., 28, *37*
Watson, D., 206, *208*
Wayne, R. P., 117, *131*
Weast, R. C., 12, *24*, 28, *37*
Weber, E., 105, *114*
Went, F. W., 143, 144, *155*
Werby, R. T., 72, 73, *75*, 102, *114*
West, P. W., 108, *114*
Westberg, K., 120, *132*, 142, *155*
Westrum, E. F., 31, *37*
Whitby, K. T., 171, 172, 174, 175, 184, 189, 190, 193, 194, 195, *209*, *210*
White, O., 143, *155*
Wiffen, R. D., 193, *209*
Wilson, K. W., 120, *132*, 142, *155*
Winefordner, J. D., 87, 88, *99*

Y

Yamamura, S. S., 110, *113*
Yamada, V. M., 40, *61*

Z

Zacha, K. E., 87, 88, *99*
Zdrojewski, A., 86, 90, *98*, *99*
Ziegler, C. S., 71, 73, *75*

SUBJECT INDEX

Numbers in italics refer to Appendix pages.

A

Absorbance, 32, *219*
Absorption by aerosols, 189, 190
Absorption band, *219*
Absorption coefficient, 32, *213*
Absorption line, *220*
Absorptivity, 32
Acid–base reactions, 106
Acrolein, 148, 154
Activation energy, 33
Activity, *220*
Adsorption, 43
Advection, 13, *220*
Aerosols, 157–207, *see also* Particulate matter
  absorption, 189, 190
  background levels, 199
  coagulation, 182–184
  composition of, 149
  diffusion, 181–184
  extinction of radiation, 185–189
  humidity effects, 204
  measurement of, 192–207
    angular integrating nephelometer, 195–206
    flame scintillation, 207
    ion mobility, 193, 194
    polar nephelometer, 206
    single particle counters, 194, 195
  particle size distributions, 165–176
  photochemical smog, 148, 149
  properties, mechanical, 176–181
  properties, optical, 184–191
  Raoult's law, 204
  size parameter, 188
  sources, table of, 163
Albedo, *220*
Aldehydes, 148, 154
Aliphatic compounds, *220*
Ammonia, 105, 106
  global aspects, 122, 123
  sources, 116
Analysis, 76–97
Angstrom's exponent, 191, 200
Angular scattering diagram, 186, 187
Angular integrating nephelometer, 195–206
Anthropogenic, *220*
Aromatic compounds, 147, *220*
Arrhenius equation, 33
Atmosphere, *220*
  circulation, 138
  color of, 201–204
  composition of, 2–6
  evolution of, 2–6
  mass of, 102
  selected data, 102
Atomic spectrometry, 86–88
Attenuation, *220*
Avogadro's constant, 27
Azo dye, *220*

B

Beer–Lambert–Bouguer Law, 32, 85
Bimodal distribution, 175, *220*
Biosphere, *220*

236

Blue moon, *220*
Box model, 17, 18
Brownian motion, 13, 48, 181–184, *221*
Bubblers, 44, 45
Budget
    chemical cycles 7–10
    sulfur 100–105
    nitrogen 122, 123
    carbon 135, 138

C

Calibration, 57–59
Carbon-14, 137, 139
Carbon compounds, 133–154
    analysis, 150–154
Carbon cycle 135, 138
Carbon dioxide
    analysis, 150–153
    effects of, 141
    global aspects, 134–139
    secular increase, 136
    solubility in sea water, 137
    sources and sinks, 134–139
Carbon monoxide
    analysis, 150–154
    from automobiles, 139, 140
    effects of, 141
    photochemical smog, 120, 121, 142
    sources and sinks, 139–142
    stratosphere, 142
Carbonic acid dissociation constants, 107
Carboxyhemoglobin, 141
Cascade impactor, *221*
Chemical kinetics, 32–36
Chemiluminescence, 128–129, *221*
Chromatography, 76–81, *221*
    column, 81
    gas, 78–80
    thin layer, 80, 81
Cl^- in rain, 72
Cloud, *221*
Cloud condensation nuclei, *213*
Cloud processes, 11, 19–22
Coagulation, 172, 182–184, *221*
Coefficient of haze, *213*
Cold trap, 42, 43
Collection, 38–59
    gases, 42–46
    particles, 46–52

Colorimetry, *221*
Column chromatography, 81
Concentration
    probability plots, 69, 70
    rate of change, 10
    wind direction dependence, 71
Condensation nuclei, *213*
Conductimetry, *221*
Constituents, stable and variable, 7
Continuity, *see* Mass continuity
Correlation coefficient, *221*
Correlation spectrometry, 96, 97
Coulometry, *222*
Cross section, scattering, 188
Cryogenic, *222*
Cunningham slip correction, 177, *222*
Cycles, *see* Budgets
Cyclic compound, *222*

D

Dark reaction, *222*
Data
    presentation of, 67–73
    treatment of, 62–74
Deliquescence, 8, 9, 19, 29, *222*
Deposition velocity, *222*
Detection limit, 64, 65
Diffraction, *222*
Diffusion, 48, *222*
    of aerosols, 181–184
    eddy, 12–17, *222*
    molecular, 11, 12
Diffusion coefficients, table of, 12
Diffusiophoresis, 52, *222*
Dimethyl mercury, 150
Dispersion
    box model, 17, 18
    eddy, 12–17
    equations, 14–18
    isotropic, *224*
    nonisotropic, 15
Drag force, 176–178
Droplets, properties of, 28, 29, 159
Dry adiabatic lapse rate, *222*
Dustfall, *213*, *222*
Dynamic meteorology, *222*

E

Eddy dispersion, 14–18, *223*
Efficiency
  of gas collection, 46
  of particle collection, 48–50
Efflorescence, 8, 9, 19, *223*
Einstein, *223*
Electrometry, *223*
Electron capture detector, 79
Electrostatic precipitators, 51, 52
Enthalpy, 30, 31
Entropy, 30, 31
Eon, *223*
Equilibrium constant, defined, 29
Error
  random, 63–65, *223*
  sources of, 63
  standard, 63
  systematic, 62, 63, *223*
Escape velocity, 3
Ethyl mercaptan, 149
Excess volatiles, 3
Extinction, 31, 32, *223*
  by aerosols, 185–191
Extinction coefficient, 31, *213, 223*
Extraction, 42–46, *223*

F

Fallout, 161, *223*
Fick's Law, 11, 181, 182
Filters, 46–52
  membrane, 48–51
  Nuclepore, 49–51
  selected data on, 49
Flame ionization, 78, 153
Flame photometry, 111, 112
Flame scintillation, 207
Flowrate measurement, 53–55
Fluorometry, 88–90, *223*
Flux density, 12, *223*
Fog, *223*
Formaldehyde analysis, 154
Fossil fuel, 137, 138, *223*
Free energy, 30, 31
  of formation, table of, 31

Freon, 150
Fugacity, *223*

G

G, see Free energy
Gas chromatography, 78–80, 153, 154
Gases, properties of, 26, 27
Gaussian distribution function, *223*
Geometric mean, 169
Gibbs free energy 30, 31
Grab samples, 40, *224*
Gravimetric methods, *224*
Greenburg–Smith impinger, 44, 45

H

H, see Enthalpy
Haze, *224*
Hemoglobin, 141
Henry's Law, 27, 28
  effect on carbon dioxide, 137
Heterogeneous nucleation, 159
Homogeneous nucleation, 159
Hydrocarbons, *224*
  ambient concentrations, 146, 147
  analysis, 78, 153, 154
  anthropogenic, 144–147
Hydrogen peroxide method for sulfur
  dioxide, 109, 110
Hydrogen sulfide analysis, 111
Hydrometeor, *224*

I

Ideal gas, 26, *224*
Illuminance, *224*
Impaction, 48, 162, *224*
Impactors, 51, 181
Impingers, 51
Imprecision, *224*
Infrared, nondispersive, 150–153
Infrared spectrometry, 90, 91
Inlet tubing, 40
In situ, 224
Intensity, *224*
Interception, 48, *224*
Inversion, *224*

Ion mobility, 193, 194
Isopleths, 73
Isotropic, *224*

J

Junge distribution, 174

K

Kapteyn's rule, 171
Kelvin effect, 28, 29, 159
Kinetics, 32–36, *224*
Knudsen number, 176
Köhler curves, 160
Koschmieder theory, 196, 199

L

Laplace's equation, 11
Lasers, 95, 96
Lidar, 39, *see also* Lasers
Limit of detection, 64, 65
  table of values, 87
Lithometeor, *225*
Lithosphere, *225*
Log-normal distribution, 70, 169, 174, *225*
Luminance, *225*

M

Mass continuity, 11, 22
Mass concentration, *213*
  and light scattering, 199
Mass distribution, 167
Mass spectrometry, 93–95
Mean, *225*
Mean free path, *225*
Mercaptans, 149, 154
Mercury compounds, 150
Mesoscale, *225*
Metal organic compounds, 150
Meteorological range, 196, *225*
Methane, 142, 143
Methods of analysis, generalized, 38, 39
Methyl mercaptan, 149
Methylene blue method for reduced sulfur,
  111

Micrograms per cubic meter, defined, 56,
  213
Micrometeorology, *225*
Midget impinger, 44, 45
Mie scattering, 96, 185, *226*
Mixing ratio, *212*, *226*
Mobility, 178, *226*
Mode, *226*
Molar absorptivity, *226*
Molar extinction coefficient, *226*
Molarity, *213*
Mole, 26, *226*
Mole fraction, 27, *212*
Molecular concentration, *213*

N

NDIR, 150–153
Nephelometer, 195–206
Nephelometry, *226*
Neutron activation analysis, 92, 93
Nitrogen compounds, 115–129
  analytical methods, 123–129
  global aspects, 122, 123
  reactions of, 115–122
Nitrogen oxides
  analysis, 124, 125, 128, 129
  reactions of, 117–122
  sources, 118, 119
Nitrous oxide, 116, 122–124
NMR, 91, 92
Nondispersive infrared, 91, 150–153
Normal distribution, 69, *226*
Nuclear methods, 92, 93
Nucleation, 159
Number distribution of particle size, 167

O

Olefinic compounds, *226*
Optical thickness, *213*
Order of reaction, 33
Oxidation, *226*
Oxygen
  evolution of, 36
  excited state, 120
Oxygen atoms, 120
Ozone, 115–129

analysis, 125–129
natural source, 121, 122
reactions of, 117–122

P

PAN, 79, 121, 149
Paraffin, *226*
Partial pressure, *213*
Particles, collection of, 46–52
Particle concentration, *213*
Particulate matter, *see also* Aerosols
  global considerations, 162–165
  sinks, 161, 162
  sources, 158–161
Parts per million, defined, 55, *212*
Permeation tubes, 58, 59
Peroxyacetylnitrate, *see* PAN
Pesticides, 150
pH of raindrop, 106–108, 112
Phase, *227*
Photochemical smog, 117–122
  secondary products, 147–149
Photochemistry, 34, 35, *227*
Photolysis, *227*
Photometry, *227*
Photophoresis, 52, *227*
Photosynthesis, *227*
α-Pinene, 143
Plume diffusion, *see* Eddy dispersion
Plume width, scale of, 16
Polar nephelometer, 206
Pollutant, *227*
Polycyclic compounds, *227*
Potassium iodide, *see* Ozone analysis
Potentiometry, *227*
Power law distribution, 174
Precipitation, 21, 23, *227*
Precision, *227*
Preexponential factor, 33
Pressure fraction, *212*

R

Radiation, extinction of, 31, 32
Radioactivity, *213*
Rainout, 21, *227*
Rainwater, composition of, 102

Raoult's Law, 27, 159
Rate of reaction, 33
Rate constant, 33
Rayleigh scattering, 184–185, *227*
Reduction, *227*
Relative humidity, *212*, *227*
Remote sensing, 95–97
Residence time
  atmospheric, 104, *228*
  instrumental, 65, 66

S

S, *see* Entropy
Saltzman method for nitrogen oxides, 124,
  125
Sampling, 38–59
  adsorption, 43
  extraction, 42–46
Sampling inlet, 40
Sampling train, 57
Saturation ratio, *228*
Scattering, *228*
Scattering angle, *228*
Scattering coefficient, 32, *213*
Scattering cross section, 188, *228*
Scattering efficiency, 188
Scavenging, *228*
Sedimentation, *228*
Selective removal, 43
Sensitivity, 64
Settling, 160
Settling velocity, 178
Single particle counters, 194, 195
Sink, *228*
Size parameter, 188
Smog, *see also* Photochemical smog, *228*
Solubility, *228*
Solutions, 27, 28, *228*
Sorption, 19, *228*
Spectral, *229*
Spectrometry, 82–97, *229*
  atomic, 86–88
  correlation, 96, 97
  infrared, 90, 91
  mass, 93–95
  nuclear magnetic resonance, 91, 92
  ultraviolet, 85, 86
  visible, 85, 86

Spot tests, 81
Standard error, 63
Standard temperature, 56
Static stability, *229*
Steady-state condition, 34
Stoichiometric proportions, *229*
Stokes's law, 176
Stokes number, 47, 48, 179–181, *229*
Stop distance, 180
Stratosphere, *229*
Suess effect, 137, *229*
Sulfur compounds, 100–112
    ambient concentrations, 101
    analysis, 108–112
    global considerations, 100–105
    organic, 149, 150
    oxidation of, 105–108
    reactions of, 105–108
    sources, 100–105
    table of sources, 101
Sulfur dioxide
    analysis, 108–112
    oxidation, 109, 110
Sulfur hexafluoride, 79
Sulfuric acid, 100
Sulfurous acid dissociation constants, 107
Supersaturation, *229*
Surface area concentration, *213*
Synoptic scale, *229*

T

Teragram, *229*
Terpenes, 143, 144, *229*
Tetraethyl lead, 150
Thermal conductivity, 78
Thermal force, *229*
Thermal precipitator, *229*
Thermodynamics, 29–31

Thermophoresis, 52, *229*
Thin-layer chromatography, 80, 81
Thorin indicator, 110
Time constant, 65, 66
Titration, *229*
Titrimetric methods, *230*
Transmittance, *230*
Transmissometer, 96
Troposphere, *230*
Turbidimetry, *230*
Turbulence, 12

U

Ultraviolet-visible spectrometry, 85, 86

V

Van't Hoff factor, 159, *230*
Vapor, *230*
Variability, index of, 6, 7
Velocity, settling, 178
Visibility, *see* Meteorological range
Visual range, *230*, *see also* Meteorological
    range
Volume concentration, *213*
Volume fraction, *212*
Volume measurement, 53–55
Volumetric methods, *230*

W

Washout, 21, *230*
Water, vapor pressure of, 28
Weight fraction, *212*
West–Gaeke Method for sulfur dioxide,
    108, 109